Glyphosate & the Swirl

Glyphosate

the

Vincanne Adams

Critical Global Health: Evidence, Efficacy, Ethnography
A series edited by Vincanne Adams & João Biehl

An Agroindustrial Chemical on the Move

Swirl

Duke University Press Durham & London 2023

© 2023 DUKE UNIVERSITY PRESS All rights reserved
Printed in the United States of America on acid-free paper ∞
Project Editor: Lisa Lawley | Designed by Aimee C. Harrison
Typeset in Portrait Text Regular and Helvetica Neue LT Std
by Westchester Publishing Services

Library of Congress Cataloging-in-Publication Data
Names: Adams, Vincanne, [date] author.
Title: Glyphosate and the swirl : an agroindustrial chemical on the move /
Vincanne Adams.
Other titles: Critical global health.
Description: Durham : Duke University Press, 2023. | Series: Critical global health |
Includes bibliographical references and index.
Identifiers: LCCN 2022033075 (print)
LCCN 2022033076 (ebook)
ISBN 9781478016755 (paperback)
ISBN 9781478019411 (hardcover)
ISBN 9781478024033 (ebook)
Subjects: LCSH: Medical anthropology. | Medical policy. | Glyphosate—Toxicology. |
Herbicides—Toxicology. | Agricultural chemicals—Social aspects.
Classification: LCC GN296 .A34 2023 (print) | LCC GN296 (ebook) |
DDC 306.4/61—dc23/eng/20220804
LC record available at https://lccn.loc.gov/2022033075
LC ebook record available at https://lccn.loc.gov/2022033076

Cover art: Søren Solkær, *Black Sun #59*. Courtesy of the artist.

To my dear friend and colleague,
Sharon R. Kaufman, whose wisdom,
devotion, and love for our work will
persist far beyond her foreshortened life

Contents

Like a swirl, this text has moved through various clusters and coalescings of the materials, landing in this way for the time being. I am indebted to colleagues who read early—in some cases, quite early—drafts and portions of this book (some of whom may not recognize this final iteration as related to those early forms). The list of people to whom I am very grateful includes Cori Hayden, Julie Livingston, Miriam Ticktin, Zoe Wool, Yuyang Mei, Sharon Kaufman, Alex Nading, Nicholas Shapiro, Raphael Frankfurter, Stacy Leigh Pigg, Emma Shaw Crane, and Rayna Rapp. I benefited from thinking about this project with audiences for talks I delivered at the Anthropology and American Studies departments at New York University; the Anthropology and Sociology Department at Simon Frasier University; the Department of Anthropology, History, and Social Medicine (now the Department of Humanities and Social Sciences) at the University of California, San Francisco; and the Helsinki Collegium of Advanced Studies at University of Helsinki, Finland. This project similarly benefited from a wonderful group of scholars who participated in a workshop called Alterlife, especially Michelle Murphy, Lochlann Jain, Hannah Landecker, Mel Chen, Kelly Knight, Alex Nading, Bishnu Ghosh, Nicholas Shapiro, Galen Joseph, Nadia Gaber, Sheyda Aboii, Nicole Mabry, and Kathryn Jackson. I also thank Rebecca Newmark, Yogi Hendlen, and Glenn Stone for insights offered on portions of the work. I want to thank Daphne Miller, MD; Michelle Perro, MD; and Michael Antoniou, PhD, for their direct and indirect contributions to my thinking on the topic. I also thank Paicines Ranch for their generous support of activities mentioned in the book. I benefited enormously from the serious and detailed suggestions from four anonymous reviewers. Finally, I want to thank Ken Wissoker, who has been a model editor, colleague, and friend for many years, and whose steadfast and acute insights about my work and our field has helped make this little academic swirl that I live in quite wonderful indeed.

From blossoms comes
this brown paper bag of peaches
we bought from the boy
at the bend in the road where we turned toward
signs painted *Peaches*.

From laden boughs, from hands,
from sweet fellowship in the bins,
comes nectar at the roadside, succulent
peaches we devour, dusty skin and all,
comes the familiar dust of summer, dust we eat.

O, to take what we love inside,
to carry within us an orchard, to eat
not only the skin, but the shade,
not only the sugar, but the days, to hold
the fruit in our hands, adore it, then bite into
the round jubilance of peach.

There are days we live
as if death were nowhere
in the background; from joy
to joy to joy, from wing to wing,
from blossom to blossom to
impossible blossom, to sweet impossible blossom.

—Li-Young Lee, "From Blossoms" (1986)

From Blossoms

The story of how I first met glyphosate begins with an encounter not unlike
that with the peach and its impossible blossoms—with things that seem too
good to last, with my recognition that something as simple and tasty as a
peach might be something about which to be worried, or that it could be
unstable in its peachiness. It happened for me during the summer of 2013 while
on a walk with my neighbor, an integrative pediatrician named Michelle Perro.
She told me that our food was causing a public health disaster.

"Our kids are in crisis," she said, as we made our way along the ridge trail
behind my home. "The kids I see are sicker than any generation before them

with chronic health problems, and these problems have become the new normal," were her words. Chronic gut issues, including irritable bowels and recurring diarrhea; chronic headaches and brain fog; persistent eczema; a host of immune system disorders, from asthma and rheumatoid arthritis to ulcerative colitis; and a cornucopia of mental health issues, from depression to hyperactivity and anxiety, were all on the rise among the children and teens she saw on a daily basis. According to Michelle, they were sick from the food they were eating, which was chock full of pesticides. Chief among the pathogenic culprits was something called *glyphosate*.

Glyphosate, she explained, is the active ingredient in Roundup, the powerful herbicide patented by the Monsanto Company. It is used in the agro-industrial production of the four major cash crops in the United States, including the genetically modified Roundup Ready crops of rapeseed (used to make canola oil), corn, soybeans, and sugar beets. This means that a good deal of the processed and packaged foods on supermarket shelves not just in the United States but globally—including breads, crackers, pastas, prepared foods made with canola oil or sugar from sugar beets—are a likely source of glyphosate. It is also used in the production of genetically modified cotton and alfalfa, which is used in animal feed for livestock and poultry that are turned into food. Glyphosate-based herbicides are used on wheat crops as a desiccant just before harvesting, she said, even though wheat has not yet been genetically designed to be grown with Roundup. Roundup is also used with other non-GE crops during fallow seasons to clear the soil of weeds before the growing season. Thus, in addition to being in food, glyphosate is in water systems, soil, parks and playgrounds, and in residential backyards when gardeners spray their beloved flower beds and walkways to get rid of unwanted plants or to eliminate poison oak and ivy. "Glyphosate is everywhere," Michelle said. Thus began my journey into a swirl of glyphosate-laden presences and possibilities.

I learned from Michelle that in the five decades since the creation of this chemical, particularly the last twenty or so years of genetically engineered (GE or GMO) crop proliferation, glyphosate had become the one of the most widely used pesticides in the world. Its pervasiveness in the lived and consumed environment meant, for Michelle, that it was seeping into not only foods and gardens but also into the soft tissues of humans where it produced a cascade of disruptions, undermining nutritional, digestive, pulmonary, neurological, and immune systems in unsuspecting kids who showed up at her clinic every day.

2

It was not just that these kids were undernourished or under too much stress, eating overly fattening foods or too much sugar, or getting too little exercise. These were the common refrains offered by her medical colleagues about the food-based health issues of children in the United States. Most of the kids Michelle saw in her clinic had plenty of food, and they often ate grains, legumes, fruits, and vegetables along with their cereals, breads, meats, chips, and sweets. They ate peaches. These children were mostly not overweight or underweight, and yet they were chronically sick. The kids she saw would limp along in partial health for years, sometimes getting worse, sometimes better, but never becoming fully healthy. By the time they got to her, these children had been through multiple therapeutic failures and been on and off antibiotics, steroids, painkillers, and psychoactive drugs. So long as they continued to eat foods laced with pesticides and glyphosate, Michelle argued, they would never get well.

This was "the GMO generation," Michelle said. These were the kids and young adults who had grown up eating this glyphosate-rich food. She offered a concerned rage about the situation, angry over the fact that so little medical science was devoted to sleuthing the health effects of foodborne toxic pesticides and frustrated about the pushback and skepticism she often received from her medical colleagues when she tried to tell them that genetically modified foods were dangerous for health because of these pesticides. She had become, in her own words, a warrior for the parents and children who she believed were fighting against an agrochemical industrial food empire that had enabled this slow poisoning to occur and a medical industry that refused to pay attention to this problem.

I was pulled into the whirlwind of Michelle's concern. Blossom to blossom, wing to wing, I came to realize that this little chemical—now saturating the planet and penetrating into soils, water, plants, foods, and bodies in swirling formations always on the move—was both pervasive and impactful. But glyphosate also presented a particular conundrum. The juries deliberating on its toxicity were many: scientists, activists, clinicians, and industry representatives had all weighed in with loud voices on this little chemical, but there was little agreement about how safe it was for humans who were likely absorbing it through their food or drinking water or from spraying it in their backyards. Michelle was convinced that claims about glyphosate being safe were not trustworthy because the agrochemical industry had paid for the research behind these claims. To be sure, it would not be the first time chemical industries had manufactured a consensus on the safety of their

3

products when used as advised. But glyphosate didn't lend itself to an easy critique of the agrochemical industry either, and not all of the claims about its safety were coming from industry.

Before long, I found myself deep in the controversy that had emerged around glyphosate. The closer I got to glyphosate and its potencies, the more I realized this chemical was, like the contested knowledge about it, not settling. The facts about it formed atmospheric clusters that coalesced sometimes in one way and other times in other ways, but never in stable ones. Like glyphosate itself, moving into plants and soils and water and guts and cells in shifting, swirling patterns of impact, the idea of stable certainty from scientific, legislative, regulatory, and activist commitments remained unsettled. Glyphosate was best understood, I realized, as a kind of swirl.

Over the next five years, I explored the kinds of evidence Michelle relied on in her clinical rounds, tracing patients' stories and also the debates about foods and pesticides. I shadowed Michelle and watched in awe as her skillful efforts to care for her pediatric clients often led to remarkable recoveries, sometimes after years of frustration and little help from mainstream doctors. I tried to make sense of the contested evidence she drew from about the harm of glyphosate and foods in which it is found. I attended retreats with organic farmers, plant biologists, and food activists. I interviewed families with very sick kids. I tagged along with food activists as they lobbied members of the California Assembly, warning legislators of the probable dangers of glyphosate.

Some of this ethnographic work resulted in a coauthored book with Michelle, who, early on, told me she had been thinking for many years about writing a book about GE foods and sick kids. I wanted to be collaborative, so I crafted a narrative that described her practice and the wealth of evidence to support her views. That book, which was truly a collaboration, was called *What's Making Our Children Sick?* It describes the limits of mainstream medical practice in relation to chronic morbidities, the frustrations of patients, and the science about harm from GE foods and their associated pesticides. The book's subtitle, *Exploring the Links between GM Foods, Glyphosate, and Gut Health*, got right to the point. We described the perfect storm that meant children's health was likely being compromised from the inside out by food that was serving up an ample dose of chemicals along with its nutrition. The links between cause and effect, we argued, could be seen only by connecting the dots between the available science about genetically engineered foods, a reluctant medical establishment, a regulatory system full of holes, and families struggling with sick kids.

4

The book was taken up with praise in alternative and integrative medical circles and by activists who were already convinced of the danger of GE foods. Michelle went on to have a robust side career as a speaker on the topic. Our book was applauded in the echo chamber. But I went on to feel the story was incomplete. Michelle wanted a book that would convincingly show how dangerous GE foods and glyphosate were, but I found myself having to set aside many of the competing facts that kept coming up as I went through the scientific material. Were we cherry-picking the facts to make our case, using only facts that Michelle insisted were reliable because they were not industry-derived? What about the facts that we pushed aside, the other ways of knowing glyphosate, that still felt to me as though they needed attention? What was going on with the science here? What was going on with chemical harm and its mysterious unaccountabilities? Over time, I was pushed toward a more complex set of anthropological questions about chemical exposure through food and about how chemicals work in bodies, in science, and in activism, and I nurtured my lingering concern over how one *could* and *should* go about making a case against a chemical when the evidence was so controversial and contested. I began to build another archive for what I realized would have to be a different book.

There were many things that troubled me about the book we wrote. For instance, I was troubled by its prioritization of children as a sentinel community for a much larger problem faced by anyone living in the ruins of agrochemical industrialism. Child health panics have, particularly in the United States, fueled a generational narrative about futures and climate demise that prioritizes a gendered labor of care, reinforcing heteronormative demands that align scientific facts with problematic inequalities (Edelman 2004; Lancaster 2011). While Michelle's work offered the spark that first interested me in chemical harm from agroindustrialism, and while most of the ethnographic materials I gathered were drawn from shadowing her, I also knew that these well-rehearsed idioms of concern over children were problematic and that I was more interested in chemical-human entanglements not necessarily invested in heteronormative reproductive futurity. I wanted to explore the larger engagement with science and politics and the potential for understanding this chemical that a broader field of inquiry would afford.

I also knew a second book would have to move not only beyond harm to children but also beyond the role that human suffering plays in relation to activism around this chemical. Glyphosate has come to hold larger-than-life potencies because of its histories, its many constituencies, its chemical

5

opportunisms, and its ability to create all kinds of relationships with science, bodies, activism, and the facts. Glyphosate has many of what Eben Kirksey (2020) calls *chemosocialities* across environments, sensations, scientific archives, and capitalist and political opportunities. How glyphosate builds its constituencies and how clinicians learn to sense the presence of glyphosate in damaged body tissues are processes that shift our focus from abject suffering to wider questions about how we live with the chemicals that have become ubiquitous in our times.

Tracing accountability for chemical harm from food, much like tracing environmental chemical harm from anything, leads to a recognition of how quickly such accountability is diffracted in the scientific literatures, histories, regulatory practices, and activist efforts that offer competing facts about them, about which social scientists have written much. So, while it is not hard to tell at least one story of glyphosate's toxicity—indeed, that is what Michelle and I did—one can also tell other stories about how we got here with this chemical and how it alters our ability to do things like trace harm and form accountabilities at all. This is not just because industry science has whitewashed the facts. It is also, I would argue, because of the ways that glyphosate has altered the way we live.

A Chemical on the Move

Competing accounts about glyphosate's safety line up in confusing ways—or one might say that they refuse to line up. For instance, just as Michelle and I were finishing our book, the International Agency for Research on Cancer in Europe (IARC) reported that glyphosate was a category 2 carcinogen. Soon after the report came out, several lawsuits were launched in California by people who were diagnosed with non-Hodgkin's lymphoma and had used large quantities of Roundup at some point in their lives (Levin 2018; Egelko 2019). These lawsuits ended with favorable decisions for the plaintiffs, and Monsanto (which holds most of the patents for glyphosate-based herbicides) was held liable for large payouts to them and their families. But this report by no means brought an end to the debate over glyphosate's toxicity, nor did it end the struggle of the plaintiffs, some now dead, for compensation. Soon after the lawsuits concluded, lawyers and scientists hired by Monsanto began to appear in public debates about glyphosate, protesting the idea that it was toxic when used safely (using the argument that "the dose makes the poison"). On public television shows and popular websites, they argued that the levels currently approved by the EPA for glyphosate use were safe. As another tactic, they noted that glyphosate's WHO classification as a category 2a carcinogen made it only as carcinogenic as "manufacturing glass, burning wood, emissions from high temperature frying, and work exposure as a hairdresser" or even drinking coffee (GMO Answers 2015).

Monsanto filed legal appeals to the court decisions, though not on grounds that the company could not be held responsible for knowledge of glyphosate's toxicity at the time the plaintiffs used Roundup (indeed, company reports maintained it was harmless to humans when it was patented as a weed killer). Nor were its legal arguments based on the idea that category 2a carcinogenesis means the toxicity is relatively benign, especially when compared to the toxicity of other pesticides. Rather, the appeals were based on the claim that the current scientific evidence used by the IARC was not, in fact, sufficient to show carcinogenicity. Researchers from the European Food and Safety Agency (EFSA, a competing EU policy making scientific organization) revisited the data in the studies used by the IARC and argued that the scientific conclusions of that report were untrustworthy (Portier et al. 2016). The US Environmental Protection Agency (EPA) weighed in, ruling that glyphosate was safe at levels currently being used. This was not the end of the story, and later I will return to the continuing arbitration over the science.

To this day, the debate over the safety of glyphosate to humans who are exposed to it at levels currently approved by the EPA remains unsettled as multiple constituencies continue to defend, impugn, or ignore its reputation. So, despite the spiraling increase in evidence-based data proving glyphosate is or is not safe, the debate has simply become more prolific, more heated, more contested, more of a swirl.

Probing why and how glyphosate was produced, I came to understand its complex origin story and how hard it would be to know for sure how safe the foods grown with it were for humans, let alone soils, ecosystems, and farmers. What it is and how it was created make it able to circulate in multiple spaces of opportunity. Its multiplicity has given rise to competing claims about its danger and safety, as have the people and constituencies who have come to know it and care about it. Indeed, the several-decades-old debate about both GE foods and their associated pesticides among concerned publics, activists, and scientists has produced a veritable ocean of skepticism about all of the facts about them, including deliberation over the ample evidence of their potential harm but equally compelling arguments about their utility and safety. There is a growing chorus of voices arguing that glyphosate is toxic, but glyphosate continues to have advocates who believe in its utility

7

for farming and its harmlessness to humans. Its advocates include scientists from many different kinds of institutions and with many different kinds of expertise. The praise for glyphosate is not just from agrochemical industries that continue to make profits on glyphosate products; it is also from farmers, organic farmers, and pro-science activists who have lumped opposition to glyphosate with the sentiment driving anti-vaxxers and science-deniers.

Glyphosate has become the poster child for a radically successful agroindustrial empire, replacing more toxic pesticides and herbicides like Agent Orange by being paired with genetically engineered crops. But glyphosate has also become a poster child for harm from GE foods—the anchor for activists and scientists—in courts of law and legislative bodies where the goal is to eliminate it altogether, along with GE foods. Watching these debates unfold and tracing glyphosate's origin story in this book, I aim to show that glyphosate is an unstable and unreliable actor despite the fact that it has radically changed worlds. This is a different argument than one that says that glyphosate's toxicity is under- or overplayed. My focus offers an engaged anthropological tracing of what happens when we try to take an activist position against chemical harm when the chemical itself is both pervasive and hard to know. It offers a glimpse of life in the swirl.

Making sense of all the competing positions on glyphosate takes more than simply carving through the data to decide where the preponderance of evidence lies. Navigating the claims about the safety and toxicity of glyphosate means getting comfortable with the refusals of clarity and certainty that we try to rely on to arbitrate these things. One must become comfortable with a bit of vertigo. This vertigo may be familiar to anyone who has tried to decipher the scientists' and activists' claims about industrially generated chemical harm authorized by policy (Murphy 2006, 2018a, 2018b, 2018c; Ofrias 2017; Nading 2015; Shapiro 2015; Chen 2011). It will also feel familiar to those whose commitments to the social-scientific analysis of "the facts" as sociomaterial constructions leave them standing on a slippery slope in their desire to harness scientific facts and produce lines of accountability in relation to chemical harm (Boudia and Jas 2014b; Liboiron et al. 2017; Frickel and Edwards 2014). The science of chemical toxicity frequently lets us down in efforts to source accountability and redress. So too does the analytical toolkit of science studies scholarship when it comes to aligning activism with science. We cannot use the scientific facts when they go our way but dismiss them when they align with a consensus we disagree with.

What we are left with in the case of glyphosate is, to press the point, a swirl of animated possibilities: knowledge, policies, activisms, cancerous growths, cellular images, and empirical facts that are often contradictory yet occupy the same space temporarily and in persistently mutating ways. Like the clustering movement of a flock of starlings in the evening sky, glyphosate's presence moves and swirls, producing ideas about its potencies that appear certain in one moment and dissipate in the next. This book endeavors to trace the multiple paths into and through scenarios and circulations of the lively presence of glyphosate as a provocateur and animated example of the swirl—a swirl that also disrupts many a conventional assumption about being able to trust a scientific consensus.

A swirl offers only temporary agreements and certainties, forming clusters of activism in one direction in one moment, then at other times forming clusters of free-market opportunities that enable glyphosate products to be sold far beyond the wastelands of agroindustrialism and deployed in backyards and parks. Like the spray of pesticide-filled glyphosate products across fields of genetically engineered crops in the United States—spray that, as Vanessa Agard-Jones (2014) notes, settles in ways that cannot be easily managed or accounted for—the swirl of chemicals that we live with seems to be ever in motion and difficult to trace even as its effects remain potent across a wide swath of material and social spaces.

No matter how much anyone might want to hold fast to the facts and materialities of glyphosate, they will be left with a dizzying sense of its ephemerality in terms of what exactly it is, despite its ubiquity. Its swirling presence can be seen at the largest scale of inquiry (in environments and agroindustrial empires) but also in the most minute places (such as the human bodies that apparently now absorb large quantities of it). There too, it has a swirl-like presence; harm in bodies may settle in tumors and mutated lymph cells, damaged guts and disrupted digestion, or systemic dysbiotic failures, but it does not do so in every body, nor all the time. Its potential mutagenicity in human lymph cells, guts, and organs—just like its deadliness to plants and its capacity to give life in genetically engineered crops—make glyphosate a powerful thing, with potencies that bend, flex, and swirl.

Unlike other accounts of the dead-ends of science in relation to chemical harm (which I'll return to later), the problem of harm from GE foods and glyphosate seemed, the closer I looked, indisputably productive in the ways that it refused to settle. Thus, as my stack of materials that didn't fit into the first book grew, I shifted toward another book that was less about sick children than the radically impaired politics of knowledge and the lack of traction this chemical engendered in dealing with chemical harm more generally. For that book, I have found glyphosate to be a singularly useful guide.

9

Glyphosate and the Swirl is about how we have taken a chemical with many different potencies and nurtured its ambivalence in and through practices of science, capitalism, regulation, and activism that have come to govern our engagements with both chemicals and the facts about them. This book is about how all of these possibilities—all of glyphosate's constituencies—are in some way co-constituted by the chemical itself in its ability to shape-shift into a biological-then-chemical-then-biological thing and by the human constituencies, institutions, and politics that have played important roles in making these possibilities form a swirl. To study glyphosate ethnographically is to be swept up in the swirl of unstable certainties that have made it what it is.

Follow the Chemical

In following glyphosate I take inspiration from George Marcus, who advised us to follow the thing in undertaking multisited ethnography (Marcus 1995), and also from Arjun Appadurai's (1986) earlier recognition that things have vivid and complex social lives, even when tangled up in commodity systems that appear to have unalterable demands for them. These efforts to focus on *the thing* have lately come into conversation with a new ontologies approach that shifts focus away from the human and human sociality and toward explorations of how material things co-constitute worlds—a tactic that also carefully avoids reproducing simplistic scientific accounts of these things while using them to "think" in new ways about these beyond-the-human encounters (Puig de la Bellacasa 2017; Murphy 2017b; Weston 2017). My interest is in following glyphosate attentively, ethnographically, through its many contacts, engagements, and ontological transformations, in order to learn something about how knowledge production and sociality operate in its wake and how to rethink the meaning of knowledge in relation to its material exigencies.

Glyphosate, because of its ubiquity and multiple potencies, has created what Karen Barad (2010), borrowing from quantum physics, refers to as a *diffraction* of its potencies as it participates in different political and material ecosystems. Indeed, it is a chemical that has created many effects as it has moved through various terrains, sometimes creating blind spots and other times forging sensationalist visibility in its relations with the agrochemical industry, farmers, scientists, activists, and juries in courts of law. The demands made upon it constantly shift and swirl, and it offers a continuous stream of data and interruptions from a cellular level all the way up to larger infrastructural and social worlds. These travels disrupt even the firmest commitments to singular modes of accountability. This multifaceted chemical, in other words, wreaks havoc with our sense of truth making, erasing distinctions between innocent and non-innocent (Ticktin 2017) scientists, industry and activist science, consensus and corrupt claims to the facts.

I also take inspiration here from Anna Tsing's (2015) *patchy anthropology* as a model for an ethnographic journey guided by one's object of study—for her a mushroom, for me a chemical. Tsing shows how various social and material infrastructures become visible as we track the matsutake from its itinerant harvesting communities in the Pacific Northwest to seller's markets in Asia and the laboratories and archives of mycologists, and the histories of capitalism that created environmental conditions ripe for this mushroom's growth. Along the way, the matsutake gives us a roadmap, an ethnographic cartography that guides us to these places in ways that are connected by the fungal landscape, much as the mushroom enables itself to live in the world. Another useful term for this sort of work is *rhizomatic*—literally, in the way mushrooms grow, and figuratively, the way that other scholars have imagined moving away from ethnographic, theoretical, and narrative linearity (Deleuze and Guattari 1987).

Glyphosate cannot provide me with a tidy linear narrative or a series of cause and effect relationships and events—only multiple unstable homes for blame and accountability in relation to its safety or harm. This makes sense, as chemical harm does not work in ways that form firm linear lines of cause and effect; chemicals don't work this way. Glyphosate offers a multiplicity of pathways to think about chemical harm in bodies. It suffuses itself through different environments and bodily sites in heterogeneous ways in laboratories, soils, plants, foods, bodies, and governmental panels. The resulting ethnography offers a patchiness that emerges from the lives of glyphosate itself, which are different in different places, though connected by the chemical.

Although Tsing's notion of plantation economies might also direct our attention to a critique of agrochemical industrialism in relation to the ethnic and racial inequalities that have come with these food systems (Reese 2019; Guthman 2019), I will not focus in this book on those most harmed by the infrastructural choices of agrochemical capitalism (farmworkers and industrial livestock workers [Holmes 2013; Blanchette 2020]), nor on the other chemicals used in agrochemical industrial farming that are considered to be much more toxic than glyphosate (Saxton 2015; Eskenazi et al. 1999; Agard-Jones 2014; Lyons 2018). Although some of the problems I trace and analytics I borrow from in order to follow glyphosate are elucidated by others who have parsed these problems, my focus has been in the other direction, among those who are not laborers of the agrochemical industrial enterprise and yet

11

still suffer from its effects, and among the metaworlds where glyphosate has caused trouble, including the regulatory, scientific, and activist worlds. My focus, in other words, is on the upstream problems—the slippery way that facts about this chemical are diffracted at the very moment that traction on accountability for its harms becomes possible. Because of this diffraction, the effects of glyphosate show up in unlikely places all the way up the social architectures of privilege that it inhabits. The upstream social and epistemological problem sites on the agricultural food system axis are partly ones that this chemical has created.

I see this work as situated in conversation with other anthropological efforts to *follow the chemicals* in their lively relations with humans, including Brett Walker's (2010) careful tracing of mercury in Japan, Michelle Murphy's (2006, 2018b) explication of PCBs as chemical kin, Nicholas Shapiro's (2015) attunements to formaldehyde, and Hannah Landecker's (2019) metabolic approach to arsenic. These authors are joined by innovative scholars who are reimagining how to live and deal with chemical injury and environmental pollutants. We need new ways of talking about life with chemicals that inspire new ways of theorizing our already chemically altered life, or *alterlife,* as Michelle Murphy (2018a) calls it, or that reveal how chemicals form, again, chemosocialities, in Kirksey's (2020) sense. These approaches ask us to take the chemical form and its capacities as a starting point for experiencing and thinking about them, including their sensorial impingements and opportunisms (Chen 2011), in their encounters in many worlds.

With glyphosate as my guide, I will trace both how we have come to live with so much of it and how, as a particular chemical, it opens spaces for thinking about chemical injury as a swirl in much the same way that the chemical moves from soils to foods to guts in swirling formations of varied parts per billion of absorption and perhaps deadliness. Glyphosate offers multiple opportunities for thinking about the reliability and utility of the facts about it, animating certain kinds of politics, activisms, and relationships to knowledge that are in constant motion. Glyphosate is, in this sense, not a fact but a set of relations and possibilities that are constantly being redistributed (Murphy 2017a). This is true for many chemicals. I am interested in this one.

To reiterate that I am offering an engaged anthropological account and not a typical science studies account of glyphosate, I do have a position about glyphosate, and my goal is to show how we have gotten into our current predicament in trying to be activist about it. I think we should be concerned about glyphosate. To make a case for this, I focus on the key moments, places, and processes that have enabled this chemical to flourish, including those tied to capitalism (particularly academic capitalism). I am invested in showing how, when we try to be activist about chemical harm, we are caught up in a swirl.

For instance, the birthplace of glyphosate as we know it in the United States is the Monsanto Company. If one wants to know how glyphosate came to be so pervasive in the world today, one must spend some time on what Monsanto is and why it cared so much about this chemical, and what kind of chemical glyphosate was made to be. Having said that, I offer the caveat that I am not neutral about the way I read this company and its scientists' activities. I am weighing in on corporate innocence. To be sure, their goals, which led them to achieve glyphosate ubiquity, may have had fits and starts and lots of uncertainty in the ways they understood and promoted this chemical (and some of this is captured in the materials I present here), but that did not prevent the company from making it seem as if the facts about its safety were certain (along with the foods they designed to be used with it). The scientific breakthroughs that made it possible to create GE foods that could withstand the spraying of glyphosate were never, in this sense, much debated, nor were the scientists apparently much concerned about caveats in their knowledge base. When questions were raised about the possible toxicity of this chemical or its use in GE foods at Monsanto or its partner companies, such concerns were mostly answered in ways that mitigated interference in corporate goals, as we will see in the next chapters.

At the same time, I am not trying to offer a simplistic tribunal against the scientists or companies of agrochemical industrialism. It is easy to be against Monsanto. It is perhaps easier to be against it now than ever before, even though it no longer exists as a company per se. My goal is simpler and riskier. I follow glyphosate as a key actor in the creation of chemically rich foods, tracking its lively relations in nonliving and living systems, in bodies and soils, in scientific archives, and in the politics, activisms, and clinical encounters it has spawned. In all these sites, glyphosate has been made to serve agrocapitalism in specific ways; it also, as a chemical, afforded certain silences and slippages around its potencies. These affordances have been taken up by constituencies who have cared about glyphosate in ways that have come to seem oppositional. No matter what one is convinced about in terms of glyphosate's toxicity, the conditions of its presence in the world now disrupt our conventional ways of being certain about harm even while those potency-derived silences and slippages continue to alter the things that glyphosate touches. My goal is to emphasize that we are in a pickle not

13

only from environmental and public health perspectives, but also in our approaches to knowledge, science, and certainty because of the presence of chemicals like glyphosate in our world.

I also take as inspirational the probability that glyphosate's swirl—in which we cannot see an easy path to living with or without it despite its ephemeral appearances in one scientific consensus or another—is a situation that we have gotten into because of the effort to solve certain problems in and through structures that are themselves harmful. I believe glyphosate has become what Julie Livingston calls an ouroboros (a snake eating its own tail): offering new opportunities for farmers, for feeding the world, for reducing chemical harm, but in ways that have likely also been damaging and, in that damage, have generated new kinds of fixes that, in turn, may be even more harmful. Glyphosate-driven agriculture has altered people and US farmland. Its novelty as a safer and more effective alternative to other herbicides distributed alongside harsh pesticides like Agent Orange made it appealing for use in the biotechnological revolution of genetically redesigning foods, but these foods may be producing wastelands of dead soils that must now be propped up with costly seeds, pesticides, fertilizers, and nutrient additives. Glyphosate may be responsible for foods carrying doses of gut-microbe-killing toxicants, and thus responsible for large numbers of people who live from one chronic ailment to the next.

All of this is contested by those who feel glyphosate remains safe and effective in agriculture, even while the company that now makes glyphosate-ready foods and Roundup products (and profits on patents from it) have begun talking about retiring it in favor of more harmful combinations and formulas of pesticides and genetically engineered products. Choices that were made long ago with a sense of urgency and moral certainty about the future created chemically partnered foods that now must be reckoned with not just in bodies but also in the social and scientific architectures that have been mobilized to ferret out evidence of chemical injury and respond.

Finally, a geographical disclaimer. Although glyphosate is a global chemical, as is the knowledge produced about it, my ethnographic materials and concerns are almost entirely specific to the United States. Some of the insights I talk about here may be useful in other places since many activist communities work globally and in connection with one another (especially around GE foods), and I draw from European scholarship because it is so widely circulated in the United States, but I point to those places and studies in this text infrequently because I am focused mostly on the situation that is unfolding in the United States.

Glyphosate has traveled in ways that have given rise to many different activities and mobilized all kinds of certainties about what it is and what it can do. The forms of reason that we have come to rely on to adjudicate glyphosate in our world are diffracted and rendered multiple by this multi-potent chemical. In this sense, glyphosate has changed my understanding of both what happens and what matters in efforts to deal with chemical harm. Glyphosate, in other words, is an exemplary chemical through which to understand the formation of competing certainties around chemical harm, disrupting and persistently displacing the scientific consensus in and through what I am calling *the swirl*. I think that the lessons from the United States are also helpful to those deliberating its presence in other places.

Glyphosate has demanded certain kinds of doublethink, ambivalence, multiplicity, certainty, and uncertainty in its many relations with food, environments, and bodies. Recognizing this invites more thinking about how, as a ubiquitous chemical, it continues to shape my engagement with it as an anthropological interlocutor that shakes up certainty about facts. To think beyond certainty means to think imaginatively about different ways of reckoning life with and theories of chemical harm. Glyphosate invites me to move beyond the politics of knowledge (in which all facts are situated) to cultivate new languages for talking about chemical harm by shifting focus away from chemicals' constructed qualities and toward their material capacities to alter life. Like others, I am continually being swept up in, swirling, and shifting certainty in ever mutable ways. Living with toxic chemicals may require cultivating skill in the arts of ephemerality—of keeping knowledge trembling and acknowledging it as always partial, uncertain, and unstable—while still holding onto its material and actionable effects.

If, at this point, dear reader, you feel that you are looping and circling and settling and unsettling with my repetitions of the sense of movement of the things I am trying to say, then I have done my job. Welcome to the swirl.

To begin our journey with glyphosate, I turn to one of its origin stories in the next chapter. I start with how this chemical became a historical object that solved certain problems even while creating others. Glyphosate-resistant foods were products of agrocapitalist investment—peculiar objects that crossed and blurred boundaries between biology and chemistry, viral genetics and information systems—and these investments required certainties about them to be forged in order for them to "live" in human ecosystems. That story takes us into the company that has had the most invested in its success: Monsanto.

15

2

To somatically apprehend formaldehyde exposure means to begin apprehending the costs of late industrial infrastructures, economies, and standards of living. — Nicholas Shapiro, "Attuning to the Chemosphere" (2015)

Building the Food Chemosphere

Glyphosate was born in 1950 in the laboratory of the Swiss chemist Henri Martin, whose experiments in synthesizing new chemicals for the pharmaceutical company Cilag led him to create N-(phosphonomethyl)glycine,

a phosphonomethyl derivative of the amino acid glycine, an odorless white crystalline solid composed of one basic amino function and three ionizable acidic sites (Dill et al. 2010). He called it *glyphosate*. Because glyphosate had no apparent pharmaceutical use, Martin neither published nor patented it, but the journey of this newly born chemical did not end there. In 1959 Cilag was acquired by Johnson and Johnson Pharmaceuticals, which sold glyphosate among its chemical samples to Aldrich Chemical, which sold its samples to various companies in the 1960s. One of these was probably the Stauffer Chemical Company, which discovered that, as a phosphonate, glyphosate would bind with calcium, magnesium, manganese, copper, and zinc, enabling them to be removed from metal surfaces. Stauffer patented glyphosate in 1964 as a metal chelator useful for cleaning pipes (Komives and Schroder 2016).

Popular accounts of glyphosate's origin story also recount how, as news spread of glyphosate's ability to kill plants in areas where it was being used to clean pipes, Monsanto, a chemical company that had built a small fortune in marketing DDT (an insecticide) and Agent Orange (an herbicide), heard about this new chemical weed killer and bought the chemical. But this account is contested by others (Dill et al. 2010) who say that, independent of this discovery, the Monsanto scientist John Franz discovered glyphosate while trying to synthesize a water-softening agent (that is, a chelator) that could be, at the company's request, repurposed as an herbicide. It was one of three chemicals he was working on for this purpose. In this account, Franz describes experimenting with various analogs and derivatives of water softeners to no avail; he "was ready to drop the project. But then I began trying to figure out the peculiarities of [these] compounds, and I wondered if they might metabolize differently in the plants than the others. . . . I began to write out metabolites. . . . You could write a list of about seven or eight. . . . It involved completely new chemistry. Glyphosate was the third one I made" (Halter 2007, quoted in Dill et al. 2010, 2).

Franz discovered that glyphosate, and particularly its metabolite, or breakdown product, AMPA, worked effectively to kill plants by blocking an enzyme (ESPS) used by the plant in the shikimate pathway. The shikimate pathway is the metabolic process used by plants, bacteria, fungi, and algae to make the amino acids tyrosine, tryptophan, and phenylalanine. Without these aromatic amino acids, the synthesis of proteins is blocked, and the plant dies. In turning the chemical glyphosate into a biologically active entity—a "completely new chemistry" based on its effects on organic matter (in this case the artifacts produced by the plant's attempt to break down the chemical)—Franz was able to materialize glyphosate as if it were a whole new chemical. This, of course, was the turning point for the birth of what would become Monsanto's number one broad-spectrum weed-killing formula: Roundup.

At this point, I want to suggest that the bifurcated origin story of glyphosate offers a nice introduction to the swirling qualities of the chemical itself, which was one thing in one context and another thing in a different context, merging into a shared story at the point of what it can do: kill weeds. The ambiguity or duality of its provenance is replicated in what we will see are its utilities and its capacity to play a key role in the creation of whole new formations of scientific research and product development. The story of how it has come to flourish is a circuitous one wherein such things as chemical opportunisms, risks to the public, political pushback and activism, and mergers between academia and industry play important roles in creating the permeating and ubiquitous whirlwind that has become glyphosate in the United States (and the world). The manifold exigencies that lead to glyphosate's partnership with food is one full of diversions and sidetracks that cleave from the main story but then circle back to form a coherent set of threads that can be followed to explain, in part, how we got to where we are today.

To understand Monsanto's commitment to finding substitutes for other products that had become politically problematic (DDT and Agent Orange), why it was so invested in research on weed killers, and how it ultimately used Roundup to create weed-killer-resistant foods, I will start by tracking back through more of the company's history and its turn from investments in chemical products to agricultural seeds—a journey that will take us to the birth of Monsanto. As we will see, its chemical products crossed over and blurred boundaries between the chemical industrial system and food system, ultimately producing agricultural chemicals that could be companioned with food by changing the genetic structures of plants.

Monsanto and the Journey from Petrochemical to Agrochemical Foods

Monsanto emerged in 1901 when a Johns Hopkins University scientist noticed that his work on coal-tar derivatives left a sweet-tasting residue on his arms and sold his discovery to a man named John Queeny, whose wife's maiden name was Monsanto. This fossil fuel chemical was marketed as saccharin—a high-potency sweetener that was heat resistant and would not react chemically with other food ingredients. It was the new chemical company's first commercial product (*Los Angeles Times* 2010; Reuters 2009). The company's efforts to use chemistry to solve important challenges in new market spaces, including war efforts, and to capitalize on the discovery of new and exciting chemical formulas, no matter how or where they might be used, were consistent and aggressively linked to efforts to assure consumers that their products were safe, despite sometimes being known to be deadly to living things (Tokar 1998). This was even true for saccharin, which, despite its slightly metallic aftertaste and later evidence (disputed by Monsanto) that it might be implicated in bladder cancer (Council on Scientific Affairs 1985), became a key ingredient in Coca-Cola and was not displaced in the sugar substitute market until cyclamates were created nearly fifty years later.

Over the twentieth century, high-profit Monsanto products offered chemical innovations to technology and food enrichment. These included caffeine and vanillin and nonfood products such as polystyrene, PCBs, a variety of pharmaceuticals, a long list of pesticides, and, eventually, fertilizers and seeds. Like other successful corporations, Monsanto regularly bought up companies that competed with it for market share or that developed products that fell within its purview (Bravo 2014). It frequently calved new companies by consolidating and dividing off units that contained newly absorbed competitors; some of these were companies that produced pharmaceutical products, chemicals, and chemically companioned agricultural products (Gray 2016). As I began to write this book, Monsanto was being sold to Bayer pharmaceutical company; its name was being retired.

The many permutations of Monsanto's corporate life are not surprising. Chemistry enabled Monsanto to attempt multiple crossings-over between toxic chemical and edible things because it allowed scientists to treat all of them as ontologically similar. Chemistry was what all living and nonliving things-cum-products had in common. Maintaining a vision of a world that was fundamentally interconnected and interchangeable by way of its chemical substrates would become important when Monsanto grew its agroindustrial divisions and, later, when it jumped into the life sciences of genetics to produce its most successful product lines: fertilizers and seeds meant to be grown with or as pesticides (Gray 2016).

As I said, from the beginning, Monsanto had to master the delicate dance of assuring consumers of the safety of their products, especially of the partnerships they were creating between chemicals and agricultural crops. The management of knowledge about safety becomes important to the story of how our main interlocutor, glyphosate, came to be so prolific in the world. To get to that, however, we need to return to the progression of Monsanto's corporate growth and how it moved its high-profit war chemicals into the American heartland.

Monsanto played an important role in producing chemicals for various war efforts. These included defoliating herbicides, insecticides, larvicides, nematicides, and fumigants (Guthman 2019, 80–87). In 1929, "farmers sprayed nearly sixty million pounds (27,215,542.2 kg) of the two most popular insecticides (containing calcium arsenate and lead arsenate) on crops" (Davis 2014, 13). Most of the research done on pesticides at the companies producing them was originally tied to their use in warfare but later focused on their effectiveness in relation to increasing yield and killing pests and the financial benefit to farmers using them. In the post–World War II period, many of the products Monsanto designed for use in war, including DDT and Agent Orange, became again legalized for sale in the US market, prompting Monsanto to create its agricultural division in 1960 (Bayer Group). Research

19

determining the possible side effects of these chemicals in people and environments was limited, though extant. Even in the earliest forums of deliberation about the toxicity of pesticides, scientists argued over these chemicals' utility in relation to their risk. It is worth spending some time on these stories because they replicate so well the current situation with glyphosate.

In the interwar period, physicians in the United States began to see evidence of links between exposure to agricultural chemicals and morbidities in the form of stomach upset and skin rashes, but few cases of exposure resulted in acute poisoning and death (Davis 2014). During the late 1930s the FDA created a pharmacology division and developed its signature dose-response experimental animal research platform for determining the toxicity of pharmaceutical products (Davis 2014, 37), but the regulation of pesticides was initially shunted to the Public Health Service and the National Institutes of Health. These agencies surveyed farmers about their health but offered no studies of long-term effects or experimental research on dose responses to pesticides. In 1938, the FDA passed the Food, Drug, and Cosmetic Act enabling federal agencies to pay attention to the possible toxic effects of pesticides (Davis 2014, 37), but it would not be until the EPA was formed in 1970 that pesticide toxicity would become an agenda for regulation, in part as a result of the debate over DDT.

The story of DDT is singularly instructive. It was an insecticide developed in the 1940s to combat malaria, typhus, and other insect-borne diseases in both world wars, and Monsanto helped to produce its chemical ingredients and distribute the product. Many companies that sold DDT were looking into ways of marketing DDT to American farmers after World War II, as they did after World War I, to help them prevent insect-borne human diseases as well as insect-borne crop pests (Conis 2010). To this end, companies, including Monsanto, not only sold it as an insecticide to farmers, but also developed a line of DDT products that included DDT-filled paints, plaster, and construction materials to make homes, yards, and walls insect-proof, expanding this chemical's use far beyond the farming market (Conis 2016). This all came to an end, after much scientific debate over many years, when reports surfaced of severe health effects from the chemical formula of DDT, especially among those who lived in areas where it was regularly used. Testimony like that from a physician for the mother of the Plyler family in Claxton, Georgia, in the 1950s began to circulate. The physician described the mother, and then the whole family as well as their chickens and livestock, developing severe symptoms that coincided with the spraying season:

She [Mrs. Plyler] had developed severe sores in her mouth and throat and a persistent "irritation" of the head that didn't give way until fall. [Her] husband "B.C." and daughters Betty and Martha suffered similarly. Before long, she noted that their symptoms were seasonal: they set in when the "big land owners" who owned the fields surrounding her home began "spraying" their crops at the end of winter, and they let up when the spraying season ended in October. The sprays and dusts coated everything on the Plyler farm: cow pasture, vegetable garden, fruit trees, chicken coops, open feed boxes and water pans, clothes and bedding hung out on the line, and everything and everyone in the house if the windows were open when the spraying began. Moreover, the Plylers' chickens were dying, and their livestock were sick, too. (Conis 2016, n.p.)

The amazing thing about the toxicity of DDT is that knowledge and uncertainty about the health effects of DDT were being researched all through the 1940s and '50s in animals in laboratories and the wild, and even though many produced ample evidence of harm in both long-term low-dose exposure and acute high-dose exposures, this research led only to the conclusion that "there was enough ambiguity in their findings . . . to suggest that DDT used judiciously did not pose a great threat to mammals" (Davis 2014, 55). All this is to say that the painstaking and plentiful research on DDT toxicity did not deter manufacturers from making or selling it so long as there remained minimal regulatory obstacles to its continued use based on safe dose modeling.

By the late 1950s, many scientists who worked with industry were adamant that the benefits of DDT outweighed the risks posed to humans. One of these, a biologist named Thomas Jukes, argued that all chemicals are potentially poisonous and useful to humans at the same time. What mattered, he said, following a familiar toxicology logic used by researchers since the 1930s to limit regulations, was the "safe dosage":

One of the oldest principles in toxicology was stated by Paracelsus almost 500 years ago: "Everything is poisonous, yet nothing is poisonous." This is quite familiar to biochemists, who recognize that several chemical elements commonly regarded as poisonous are essential in small amounts to life. Examples of these are copper, chromium, manganese and selenium. The last named of these is also carcinogenic (i.e., tends to produce cancer). Traces of practically all the elements can be detected by spectroscopic tests in most biological materials, and all living creatures contain radioactive carbon and radioactive potassium. The crucial matter

21

is the quantity of such substances that we consume in proportion to the amount that is toxic. It is comparatively easy to poison animals with table salt in high dosage. (Jukes 1971, 538)

Jukes further argued that, in weighing all the evidence, one had to conclude that not only was DDT *benefiting* the living organisms of the ecosystem (from nontargeted animals like birds to mammals) but its continued use was in fact essential for *saving* them (Clement 1972). (Jukes is also credited with discovering that feeding small doses of antibiotics to livestock fattens them swiftly.)

By the early 1960s, concern about the harms of large-scale defoliants and pesticides to environmental health emerged alongside activism against the Vietnam War. Scientists who came down opposite Jukes began to raise their voices. The field biologist and chemist Charles Wurster argued that DDT was not only *not benefiting* living organisms, but that human survival was ultimately put at risk by the harms DDT inflicted on the environment and its living inhabitants (Wurster 2015; Conis 2016). Wurster began his career at Monsanto as a chemist in 1959, but quickly decided on a different career path and became a robust supporter of environmental laws to protect fragile ecosystems, eventually forming the Environmental Defense Fund. His research showed harm caused by DDT to animals in the wild, specifically robins, starting a robust public debate with Jukes in a popular science journal's letters to the editor (Wurster 1968).

While arguments over the safety of DDT were in full swing between companies and activists pleading with regulatory agencies (a debate well documented by Frederick Rowe Davis [2014]), Rachel Carson published her famous book, *Silent Spring*, in 1962. Although she was concerned with a variety of toxic pesticides, especially organophosphates, her book was a singular force in propelling efforts to ban DDT altogether (Conis 2010; Davis 2014). With convincing prose and emotional urgency she argued that the ecological demise and cascading impacts of harm from DDT would move up the ecological food chain, setting off a chain reaction of animal and insect extinction that would eventually put humans at risk. "Synthesizing research from toxicologists, ecologists and doctors," Carson introduced a variety of new concepts in her testimony before the US Committee on Interagency Cooperation: "bioaccumulation, lipofelicity (the bonding of chemicals to fats), the passing of chemicals to offspring via breastmilk," and a new set of considerations around genetic toxicity (Davis 2014, 215, 158). She made it clear that toxic effects could happen through direct exposure but also by consuming foods grown with these chemicals, spreading the zone of impact far beyond farmers and farmworkers (Carson 1962; Davis 2014).

A great deal of effort was expended at the committee hearing aimed at quelling concerns about pesticides. In addition to the argument that pesticides were useful, if not essential, to a healthy agricultural system, advocates presented the idea that chemical companies were already acting responsibly to protect the public. One Monsanto representative even made the case for shifting responsibility for chemical harm to regulatory agencies on grounds that chemical companies were already providing ample evidence of the safety of their products when used correctly. In a classic example of diffracting responsibility by obfuscation, this senior scientist at Monsanto said, "We know a tremendous amount about pesticides in general, but I would worry about how little we know in general about food that we eat. What is the chronic effect of eating some given vegetable, say, over a lifetime period?" (Davis 2014, 178). This tactic of diverting attention away from the toxicity of chemicals in foods to the prospect that any and all foods might be toxic when eaten over a lifetime would reappear in nearly every deliberation about the safety of Monsanto agrochemicals going forward.

Carson herself endured numerous attacks against her work as a scientist for the provocative stance she took on pesticides (Smith 2001). Nevertheless, the public outcry her book aroused led ultimately to a successful campaign to ban DDT globally. The ban, which took effect in 1972, did not mean that the world was rid of the chemical, only that it could not be used as a pesticide in some places. As Balayannis (2020) has shown, stockpiles of DDT were displaced from the United States to poorer overseas locales such as Tanzania. Moreover, the banning of DDT did not mean that other—in many cases more toxic—chemicals would be eliminated from the agroindustrial food system. On the contrary, banning DDT led to an increase in the use of the more toxic organophosphates—chemicals that, as mentioned, Rachel Carson was even more concerned about than DDT due to their potential for endocrine disruption (Davis 2014, 218). These are mostly still in use today. In the end, the case of DDT offered a template for how chemical companies could manage the adjudication and legal deliberation over the harms their chemicals posed to a public that had suddenly become aware of their risk.

Public outcry over another Monsanto product, the powerful herbicide Agent Orange, also emerged around this time (Hough 1998). This chemical toxicant containing dioxin was used during the Vietnam War along with other strong herb and insect killers, including atrazine-based products, as a defoliant to eradicate trees, bushes, and even rice fields. The indiscriminate

23

spraying of this chemical formula was thought to give American and South Vietnamese forces an advantage by exposing Viet Cong villages and killing some of their food crops. However, as news of these defoliants' harmful side effects on Vietnamese families spread at the height of the Vietnam War (in part spawned by the efforts of scientists who invented them), public protest increased (Zierler 2011). Agent Orange was eventually banned in 1971, though not because of any new organizational or scientific recognition of its damaging effects on living things. Indeed, its deadliness to living things was the reason it was effective and thus perceived to be useful. Rather, as with DDT, public outcry against it led to it being banned (Robin 2010).

I digress into the story of these pesticides and the public outcry against them to illuminate the outsized role of publicity and activism in efforts to ban toxic pesticides. In the face of public opposition to its most important commercial products, Monsanto was forced in the 1960s and '70s to invest significant resources in producing scientific evidence that these chemicals were safe for use at home and in running public relations campaigns to win over consumers who were increasingly concerned about the dangers of their chemicals. Ultimately, Monsanto gave up on DDT—a move that fortuitously prompted research into other forms of pest control using genetic recombination of plants.

In the mid-1960s, Monsanto launched a public effort to portray the company as both defender of the environment and friend of farmers. These emerged in direct response to growing public fears over toxic exposures and antagonism toward chemical companies that were helping the war effort amid growing opposition to the war itself. To this end, Monsanto rebranded its war defoliants, calling them herbicides instead of defoliants and giving them Western-themed names: the herbicide Ramrod was marketed in 1964, followed in 1968 by the herbicide Lasso (both atrazine-based formulas). Recognizing that it was not just farmers who needed to be convinced of the safety of their chemicals, Monsanto even built a ride at Disneyland in 1967 called Adventure through Inner Space whose primary purpose was to convey the idea of better living through chemistry (Strodder 2017)—an effect that I recall it having on me as a child.[1] The ride represented a journey to the nucleus of a water molecule, revealing that chemistry formed the basic building blocks of all living and nonliving things. On debarkation, riders were led through dioramas featuring the molecular miracles from Monsanto: pesticides that promised healthier gardens and more plentiful foods and a series of household products, from polystyrene bowls to nylon clothing, made audible in a theme song called "Miracles from Molecules."

24

Along with publicity campaigns in the United States, Monsanto also adopted the marketing messages that were used to launch the Green Revolution (the USAID-led campaign to improve the agricultural systems of developing countries in ways that also made them customers for US corporations selling seeds, fertilizers, and pesticides), once again shifting petrochemical-derived war products to agricultural markets and keeping them dependent on fossil fuel companies (Shiva 2000; Church 2005). Here, Monsanto used the marketing language of Malthusian futurity (a rhetoric that would continue at the company into the new millennium)—that the agrochemical revolution was necessary in order to ensure the world's logarithmically growing population could be fed (Simanis 2001). Farmers the world over were told they could help to solve the future world hunger problem and share in the profits of Monsanto's agrochemical investments. Both of these promises would eventually be contested not just in relation to the disappearing food supply but also in relation to their financial risks and losses. But as chemicals desired for their deadly capacity to kill were rebranded as vital components of the mission to save humanity in and through agricultural revolutions, Monsanto worked on staying one step ahead of its critics.

It was in part Monsanto's responsiveness to public fears over the deadliness of potent chemical pesticides like dioxin and DDT that eventually led the company to develop a cornucopia of other strategies to create what they promoted as pesticide-rich nontoxic foods. At last, we return to glyphosate.

Glyphosate

After Monsanto scientists either bought or invented glyphosate (depending on whose histories we believe), it circulated through various corporate transactions that put it to different uses, enabling it to be patented as something altogether new at each step. By the logics of competitive patent law and chemical innovation, glyphosate's existence as a commodity chemical could remain multiple—that is, a chemical with multiple ontologies and origin stories, a new chemical at each turn with new opportunities for ownership and profits. Because its patented life as a chelator was not based on its inorganic chemical potential, glyphosate was able to be transformed into a patent for Monsanto as a new and newly biologically productive entity: a broad-spectrum weed killer.

Crossing over from its life as an inorganic chelator to its organic life as a weed killer, glyphosate was, by the 1970s, a "molecule [that] advanced through the greenhouse screens and field testing system rapidly and was first

25

introduced" by Monsanto as Roundup, in keeping with the company's marketing strategies aimed at appealing to farmers and cowboys in the American Midwest (Dill et al. 2010). Roundup was a formulation: glyphosate plus a surfactant called polyoxyethyleneamine or polyethoxylated tallow amine (both abbreviated POEA). POEA was designed to break down the waxy cellular surfaces of plants and enable the glyphosate to be absorbed. Glyphosate's potency in killing plants was its ability to disrupt the shikimate pathway, as mentioned, a key process of photosynthesis by which amino acids are produced. One of the best things about glyphosate was that it was thought to be harmless to animals, including humans, because human and animal cells do not have a shikimate pathway (which is also why humans and animals must rely on their diet and microbes for obtaining amino acids). Monsanto told regulators this directly, stating that "glyphosate is only toxic to plants but not to other living species, including mammals" (Gilham 2017, 45).

Monsanto histories describe this new weed killer as perplexing to Monsanto marketers. They were accustomed to selling herbicides that were selective, meaning they killed certain weeds but left crops unharmed. Glyphosate was nonselective. This meant it would be risky for farmers if the herbicide contaminated food crops.

The whole chemical mechanism of Roundup was a breakthrough for farmers. Again, according to a company report,

> In 1970 most farmers believed they had no choice but to use herbicides and tilling to control weeds. At the time, most herbicides were pre-emergent, meaning they created a chemical barrier on the surface of a field and killed weeds when they sprouted through this barrier and came in contact with the herbicide. To be effective, pre-emergent herbicides had to spread when they were applied to fields, ensuring a consistent, even barrier against sprouting weeds. They also needed to stay active for a long time so they would continue to be effective after the spring rainy season. These two common traits were environmentally problematic because active pre-emergent herbicides could wash into streams and ground water, potentially affecting wildlife and fish. The original Roundup® herbicide was different, becoming one of the most environmentally friendly herbicides in the history of agriculture. (Bayer Group 2021)

26   In other words, because glyphosate could be applied once, after weeds had appeared, and it was believed to bind to soil after killing weeds but before planting crops so as not to run off into water systems, it was thought to be environmentally friendly. Because of its indiscriminate potency in plants,

however, farmers had to be very careful in using Roundup to avoid killing their investment crops along with their weeds, using it only in fallow periods and in ways that avoided spraying directly on or near crop plants themselves. When used on living crops, it was and is used only as a ripening agent (as with sugarcane) and just before harvest (as apparently with some wheat). This limitation was the downside of what scientists believed was a terrific solution to the toxicity of more problematic herbicides like Agent Orange, and it was why Monsanto scientists invested so much more research in new permutations of its function—permutations that would eventually lead one author to call glyphosate "a discovery as important for reliable global food production as penicillin for battling disease" (Koester 2017).

This, of course, is the story of how glyphosate became a key collaborator in the creation of GE foods. To tell that story, we must turn to a man who became a very good friend to glyphosate, Howard Schneiderman. This partnership between a man and a chemical helps orient us toward the ways that glyphosate enabled entirely new fields of study, partnerships between academic and industry scientists, and the redesign of agroindustrial farming forever (Leonard-Barton and Pisano 1990).

From Glyphosate to GE Food

Howard Schneiderman was hired by Monsanto in 1979. According to his biographers, he was a well-respected insect development biologist with expertise in endocrine systems at the University of California, Irvine, when the head of Monsanto lured him away from his university position by offering him something he could not refuse: a $100 million budget, an enormous sum of money for a researcher in 1979 (and today, for that matter) (Oberlander 1993; Gilbert 1994). The then chairman of Monsanto recalled his recruitment as an easy one: "If you're a red-blooded American who has chosen research as a career, and a guy comes and says, 'Do you have any good ideas for $100 million worth of research?' it's a fantastic temptation" (Schneider 1990).

Schneiderman had already established a brilliant track record of doing research and building departments at both Case Western Reserve University and University of California, Irvine, so his move to industry was surprising to some. In his autobiography, Schneiderman describes his aspiration as one of putting "all that money" to good use by doing what few universities could: marrying the fields of biology and chemistry in order to bring the wealth of resources at Monsanto into the academy and the great minds of the academy into collaborations with industry (Schneiderman 1994).[2]

27

Schneiderman understood the challenges and risks of this effort, writing extensively about the potential conflicts of interest that academy-industry alliances would bring about. Despite plentiful examples of collaboration between industry and academia in the years prior to World War II, after the war, with large government investments in academic research, making such work happen meant dealing with a good deal of skepticism. There were treacherous ethical gaps, competition, and suspicions among researchers on both sides of the collaborations, all of which could inhibit success. Schneiderman wrote that academic researchers whose work was often publicly funded feared they could not compete with corporations' large budgets and teams of researchers who could churn out publications, especially since corporations had armies of patent lawyers working to ensure access to scientific research results. Most universities at the time were not legally allowed to patent results (products or intellectual property) from publicly funded research (Schneiderman 1994). Academic researchers also worried about limits to academic freedom as commercial product lines were prioritized over what they considered to be basic science interests. Schneiderman's biggest challenge, he recalled, was in convincing his university colleagues, not his corporate ones, that their interests in basic research would be protected in such collaborations.

Schneiderman's success at Monsanto was partly built on his ethical reasoning about the benefits of cooperation. He claimed it was essential to the mission of feeding the planet and promoting human survival, using language that Monsanto adopted to publicize its products. In one of his later publications, entitled *Planetary Patriotism*, Schneiderman argued that both biotechnology and imaginative chemistry (in which he included genetic engineering) required the resources that were available only through joint work across industry and academia. Industry involvement was necessary for, as he wrote, "preserving this planet as a livable place for our great-grandchildren and their great-grandchildren" (Schneiderman and Carpenter 1990).

At Monsanto, Schneiderman worked swiftly to help the company develop a platform for working with Washington University in St. Louis to research drug development (a line of product development aided by Monsanto's acquisition, at Schneiderman's prompting, of the pharmaceutical company J. D. Searle). His success radically increased the funding available for university-based research, fueling hitherto impossible growth in both pharmaceutical and agrochemical research. Schneiderman eventually built Monsanto's life sciences division into the most successful biotech company in US history, affectionately referred to in testimonials as "the house that

Howard built" (Cook 1990), and a model for industry and university research collaborations far beyond the world of biotechnology.

Providing generous budgets to university researchers and using the logic and languages of transparency to assuage concerns over competition and ethics, Schneiderman became the grease that turned the wheels of such endeavors. The blurring of public and private investments in agricultural biotechnology in "the house that Howard built" created "a new global industry rooted in private-property-based technoscience" (Schurman and Munro 2010, 1). Many departments and institutes of agrosciences at US universities would be, from that point in time, founded and entirely funded by Monsanto, exemplifying the academic capitalism that now prevails in agricultural academic research (Slaughter and Leslie 1997; Stone 2010). These collaborations also created a near monopoly on the scientific knowledge needed to decipher safety, enabling companies to produce not only the scientific facts about their products but also the science used by regulatory agencies to regulate these products. Critics note that these arrangements penetrated the world of academic publishing, further blurring the lines between interested and disinterested research when it came to corporate interests.

Schneiderman tested the waters of collaboration with the pharmaceutical research at Washington University in St. Louis. He then turned to the effort of convincing Monsanto to pay attention to the emerging field of biotechnology, specifically to the development of recombinant DNA research, which was occurring only in university research laboratories. Since his own academic expertise was in insect developmental and plant molecular biology, he was particularly interested in the interaction between plant and insect genes (Schneider 1990). This led him to start collaborations with researchers hoping to discover the potential for transferring genes between plants and animals.

The first successful recombinant DNA research in living organisms was carried out in 1973 by researchers at the University of California, San Francisco, and Stanford University with the creation of oil-eating bacteria.[3] This was followed by the use of recombinant DNA technologies in mice in 1974, an achievement that generated significant public concern over the patenting of living, naturally occurring things. Schneiderman, seeing the potential, convinced Monsanto to invest in research and development that would take the company in a substantially new direction, using its agrochemical profits to build a specific platform for research on the uses of recombinant DNA technologies in the worlds in which Monsanto was already established: agriculture and farming.

Schneiderman's efforts paid off when his next successful collaboration, one with Genentech Corporation, resulted in 1979 in the production and commercial sale of recombinant bovine somatotropin (rBST, a hormone that speeds up growth and, in dairy cattle, milk production), one of the first genetically modified products to enter the food supply by way of livestock.[4] If chemistry was the lens through which Monsanto scientists were able to see the world as having an endless reservoir of infinitely similar and interchangeable molecules (as their ride at Disneyland argued), the turn to biology enabled researchers in "Howard's house" to think in new ways about the great molecular world of plant and animal life together in and through the genetic code (Schneider 1990; Gilbert 1994). "Through genetic engineering," Dr. Schneiderman's obituarist wrote, "the chromosomes of plants and animals could be manipulated to produce even greater profits for the company. Dr. Schneiderman said the transition was inevitable because the new technology was based on an unlimited resource common to every living organism: genes" (Cook 1990). Schneiderman's work at Monsanto built on the platform of permissibility (ethical, material, and legal) created by the genetic manipulations of the animal kingdom in order to do the same thing for the world of food crops.

Early research on the genetic modification of plants was done by a team of researchers at the University of Ghent in Belgium (Marc Van Montagu and Jozef Schell) and, separately, Mary-Dell Chilton at the Washington University in St. Louis (one of Monsanto's collaborator institutions).[5] The goal was to genetically transform plants in a variety of ways that were, from the perspective of industry and academe, promising in both the fiscal and scientific senses. Thus, the first thing researchers at Monsanto and elsewhere needed was proof of concept. Monsanto scientists succeeded in producing this when they took a well-known bacterial plant pathogen (*Agrobacterium tumefaciens*) and engineered it into a carrier organism for inserting specific transfer DNA into a plant cell. This bacterium was useful precisely because it was able to mutate plants by inserting itself into the plant's DNA.

The novelty of this science was in figuring out how to transform something that could harm a plant into something that would eliminate this perceived harm yet still modify it at the genetic level. In other words, when *A. tumefaciens* showed up in flowers naturally, it produced gall tumors because of its plasmid, creating a pestilence that expressed itself through the modification of the plant's DNA (Schell and Van Montagu 1977). Howard's research team tinkered with the tumorigenic properties of the bacterium in order to eliminate its tumor-producing effects but retain its ability to alter or mutate DNA. To use the bacteria as a carrier organism, researchers designed the bacteria

30

to contain the desired genetic traits that they wanted to see expressed in the plant (Schneider 1990). This trope—deadliness tamed enough to become useful, but not eliminated—is replicated all the way up the infrastructure of technologies that would produce glyphosate-resistant plants, as we will see.

Once researchers at Monsanto devised a way to use the bacterium's viral potential to alter plant genes, they used a "gene gun" to shoot the modified bacteria into the plant, where it would embed in random sites on the plant's DNA and create a genetically novel entity. For the researchers involved, the fact that the foreign genes landed *randomly* on the plant's genome did not matter as much as the fact that the desired trait was being expressed in the plant at all. The description of these technologies as simply mimicking or harnessing a naturally occurring bacterial function would go a long way to enabling the companies to say they were simply enhancing and capitalizing on plants' own reproductive processes. The first proof-of-concept GE plant produced in 1982 at Monsanto was an antibiotic-resistant tobacco plant.

The breakthroughs for recombinant DNA in the plant kingdom discovered at Monsanto led to a flurry of activity at multiple research locations. Like all biotech at the time, there was competition to be the first to the finish line on genetically altered food crops. The first genetically modified food that was brought to market was made not by Monsanto but by a small Californian company called Calgene that developed the Flavr Savr tomato (Martineau 2001). The Flavr Savr tomato had been modified to ripen but not soften (enabling longer transport and shelf life). Using the same genetic engineering technologies that Monsanto researchers discovered, Calgene scientists used bacterial gene-altering technologies to alter one of the tomato plant's own genes.

Although concerns were raised among regulators, and at least one of the scientists who developed the Flavr Savr tomato began publishing concerns over whether adequate safety testing had been done on them (specifically concerning the effects of random landing sites in the host's DNA), these tomatoes were brought to market in 1992 (Martineau 2001). They were available in stores until 1994, but the public was suspicious and didn't buy them. They were, in other words, a commercial flop. Nevertheless, Monsanto soon acquired Calgene and, rather than let it continue as a competitor in the market of GE food products, began to do research on a different kind of genetic alteration. The tomato produced by Calgene used a modified copy of one of its own genes; the innovations at Monsanto were in producing transgenic plants not by inverting plants' own genes but by inserting genes from other species into their DNA. Having learned from the experience with the

31

Flavr Savr tomato, these food crops would not be marketed as novel in any way when they arrived in supermarkets; there would be no labeling or other indications that they were transgenic. They were considered substantially equivalent to non-GE foods (a concept we will return to later).

The first transgenic GE plant foods to be designed at Monsanto did not involve glyphosate but rather a novel gene technology to make them insect killers, again using Howard Schneiderman's special area of expertise. This involved using the same *Agrobacterium tumefaciens* and gene-gun technology to create a plant that contained Bt (*Bacillus thuringiensis*), a naturally occurring insecticide deadly to a variety of worms and other insects in different doses and formulations. In fact, researchers at the University of Ghent and a Belgian company called Plant Genetic Systems developed a genetically modified, insect-tolerant tobacco crop using Bt as early as 1985 (Höfte et al. 1986; Vaeck et al. 1987). Over the next decade, experiments and research on the use of this technology took place in multiple places, and by 1995 Monsanto successfully developed a Bt seed line and received approval from the FDA for the commercial sale of it in a strain of corn.

Bt is a naturally occurring bacterial toxin that had been used for many years by farmers to eliminate certain insect pests. Its toxicity derived from the fact that *Bacillus thuringiensis* proteins dissolve in the high pH gut of certain insects when consumed and become active, releasing a crystal, or Cry, protein. The Cry protein inserts itself into the gut wall, essentially punching holes in the lining of the gut. This allows bacteria in the insect gut to enter its body, killing the insect with a type of septicemia (Graf 2011).

Applied externally as a spray, as farmers historically used it, Bt toxin breaks down and dissipates with exposure to sun and moisture and can be largely eliminated from a plant before it is consumed. Inserting the Bt into the DNA so that it is expressed throughout the plant turns the entire plant, including its fruits, leaves, stems, and seeds, into a deadly killer for insect pests. It cannot be washed off. Anything (or anyone) eating these foods also eats the Bt or Cry proteins.

Bt technologies enabled Monsanto to claim that the use of Bt modified seeds would reduce the need for insecticides. This was true, but only because Bt modified seed turned the entire plant into a pesticide. Still, researchers working on the development of these crops at Monsanto assumed (and many still argue) that Bt is harmless to humans based on the fact that the pH of the insect gut is higher than that of humans, meaning the insect gut is less acidic, and the toxic Bt protein would not be activated in a low pH (acidic) environment like the human gut. Critics of this pesticide food argue that genetically modified Bt plants contain a Cry protein that is preactivated, meaning it does not require the alkaline gut of the insect to be activated (Székács and Darvas 2013). Depending on what science you read, this protein may or may not be harmful to humans (Carman 2013; NASEM 2016). What is important for the story of glyphosate is how Bt genetic modification of plants opened the door for Monsanto's more important commercial endeavor: making glyphosate-tolerant plants.

At Monsanto, Schneiderman's team was interested in developing genetic modifications that included what would be called "enhancements" (like the Flavr Savr tomato, vitamin A–enriched rice or frost-resistant vegetables or berries, that is, plants with altered qualities of nutrition, flavor, size, or shape), but Monsanto's main focus was on modifications that would solve another market problem: using its petrochemical arsenal in the agricultural market to improve yields by changing the fundamental ways that crops could be grown with Monsanto pesticides and fertilizers. In fact, the so-called enhancement work at Monsanto would only ever constitute about 1 percent of its investment in genetically modified food products despite that being the rationale for claiming GE technologies will save the planet by protecting the world's food supply (Perro and Adams 2017). The motherlode for Monsanto would arrive when Howard's house figured out how to stack the Bt trait with a genetic modification that could eliminate Roundup's deadly effectiveness. That is, they would create plants that could both kill insect pests *and* withstand exposure to glyphosate. Despite these GE modifications being presented as less toxic alternatives, their distribution did not actually decrease use of many items in the war-inspired chemical arsenal available in commercial agricultural markets, but they did make a lot of money for the company.

Within a year of commercializing Bt seeds, Monsanto produced its first glyphosate-resistant seeds, patented as Roundup Ready seeds. These seeds could withstand the spraying of Roundup not just once but multiple times over the growing life of the crop (Reuters 2009). The first plants to be marketed were soybeans; Roundup Ready soybeans arrived to the US market in 1996 after being tested in China (and after Schneiderman had died). Many other products followed soon after, including sugar beets, rapeseed, corn, alfalfa, and cotton. Other companies also jumped into the GE seed production market, including Syngenta AG and Dow AgroSciences, both now major stakeholders in the success of GE agrochemical products. Archer Daniels Midland Company was another competitor that also switched from chemicals to agrochemicals and, eventually, GE foods along a similar timeline to Monsanto's.

33

What all this means is that about two and a half decades ago, the major cash crops of the United States changed significantly. They became genetically designed to be both pest and herbicide resistant or, said otherwise, they became partners with new kinds of chemical companions. Roundup was also marketed widely for use in household yards and gardens, parks, schools, public spaces, and even sidewalks in urban and suburban regions, in addition to being used in forest management through aerial spraying (Schuette 1998) and, as mentioned, for wheat desiccation (Ball et al. 2003). Monsanto continued to make money from glyphosate even after it came off patent in 2000 by licensing formulations to many glyphosate-based herbicide producers. Today, Roundup remains a number one seller for agricultural, urban, and home use. By 2019, roughly 50 percent of Monsanto profits came from its glyphosate patents and products (Hendlin 2019).

Glyphosate's Ubiquity

Using genetic engineering technologies, Monsanto was able to use glyphosate and glyphosate-tolerant seeds to produce and profit not just from pesticides and seed crops but also from the fertilizers that would be needed to replenish depleted soils (since many of its pesticides killed the microbes and animals that soils need to sequester carbon and other minerals). The growth in use of GE crops since 1996 has been stunning. Only three years after they were introduced, over 100 million acres worldwide had been planted with genetically engineered seeds. As of 2017, somewhere between 80 and 95 percent of all soybean, canola, and corn crops in the United States were genetically engineered with Bt and Roundup Ready traits. Roughly 70 percent of all packaged dry foods in the United States that do not adhere to organic guidelines have GE ingredients in them. Similarly high percentages of GE foods appear in animal feed, especially soy, corn, and alfalfa (Hetherington 2013). Roundup Ready and Bt crops are pervasive in the United States, and, as we will see, as resistance grows to Roundup, the company has developed even more potent pesticide-resistant GE crops that use even more toxic ingredients, such as dicamba-resistant plants.

Monsanto has consistently referred to the biotechnologies of agrochemical farming as a boon not just to farming but to the sustainability of the planet and its human populations. On June 17, 1998, the CEO of Monsanto at the time, Robert Shapiro, offered this insight in his speech to the Biotechnology Industry Organization in New York City: "Biotechnology represents a potentially sustainable solution to the issue not only of feeding people, but

34

of providing the economic growth that people are going to need to escape poverty. Biotechnology poses the possibility of leapfrogging the industrial revolution and moving to a post-industrial society that is not only economically attractive, but also environmentally sustainable" (Simanis 2001).

In many ways, GE food crops, particularly glyphosate-saturated ones, are a uniquely American issue in relation to the intensity of their saturation, but glyphosate is a global chemical and efforts to limit its use and presence have been robust in many places (Robinson et al. 2015; Arcuri and Hendlin 2019; Kloppenberg 2005; Stone 2017). While political activism has pushed regulatory partners in some places (the United Kingdom and Europe) to follow "precautionary principle" policies for GE foods (Ackerman-Leist 2017), very little regulation has been offered in other places. Most countries use Roundup and Roundup Ready seeds as they are used in the United States, following allowable and "no adverse effect level" guidelines, if there are regulating agencies at all. To this day, efforts to restrict the use of GE seeds and bans on the use of Roundup vary from country to country in the European and British contexts (Schurman and Munro 2010; Ackerman-Leist 2017), and globally the use of glyphosate has been steadily on the rise for twenty years.

Presently, Roundup Ready foods are grown in twenty-six countries; they are distributed in more, along with the arsenal of pesticides provided by American agrochemical companies.[6] By 2010, glyphosate formulations were licensed to 130 countries. Because of the problem of weed resistance requiring increasing amounts of glyphosate in areas where it is used, "it is the most heavily used herbicide in the world, with an annual global production volume in 2012 of more than 700,000 tons used in more than 750 different products" (IARC 2015, 394), with use increasing 130-fold in the decade between 2004 and 2014 (Landrigan and Benbrook 2015; Benbrook 2016). Despite growing concern over glyphosate's effects in communities around the world where it is heavily used (Lyons 2018; Arcuri and Hendlin 2019; Guillette et al. 1998; Hetherington 2013; Stone 2002), there is virtually no definitive science concerning how much glyphosate remains in foods by the time they get to commercial markets, although activists have tried to undertake such studies (Robinson et al. 2015; Perro and Adams 2017).

Glyphosate forms a nexus of destruction and production that keeps farmers in business and believing they are using the most chemically minimal form of weed control they can, even while these technologies also keep farmers dependent on agrochemical companies for the triple expense of seeds that must be bought each year to avoid patent infringement, pesticides, and fertilizers (Kloppenberg 2005; Shiva 2000). It also means that the soils where they grow Roundup Ready foods are drenched in glyphosate. Despite industry claims that it binds with soil and thus does not bioaccumulate and is removed from ecosystems swiftly, it is difficult to know how such accumulation relates to increases in usage over the past decades. We know that it is found in soil, air, surface water, and groundwater, and likely in food in most places where it is used. Glyphosate loops humans into its chemical relations, showing up in the United States in the breastmilk of lactating mothers and in the urine of most people at rates of one to ten micrograms per liter (Mesnage et al. 2013; Mills et al. 2017; Adams 2016; Honeycutt and Rowlands 2014).

Glyphosate, I have argued, emerged as a provocateur of a whole new kind of science, in which molecular alterations of plants would enable them to accommodate new kinds of chemical kin. These accomplishments were driven by the chemical as much as they were by scientists like Howard Schneiderman and his team at Monsanto in collaboration with university-based scholars. Glyphosate as we know it today arose in Monsanto's laboratories by way of Monsanto's own reconfiguration of the meaning of academic collaboration, which gave rise to the most successful post–World War II forms of academic capitalism and harnessed entire academic fields (not just institutions) to work on industry products. These arrangements enabled glyphosate to become a permeating chemo transgressor in the soft tissues of many living and nonliving things (Kirksey 2020), perhaps interfering with gut microbes, chelating, the production of amino acids, and the mutation of cells and genes, though we may not know this for sure. Glyphosate has become a prolific instigator of new desired and undesired collaborations that wrap agriculture scientists, farmers, farmlands, activists, and consumers into new kinds of partnerships and opportunities, transgressing sensibilities about where the boundaries between human, food, and chemical might be drawn and making it hard to decipher whether harm is being done.

The swirling, looping forces of history that merged corporate and academic interests, chemical promise and the tricky management of chemical risks, Monsanto's effort to solve agricultural problems and repurpose its products, its efforts to both evade and respond to public concerns . . . all of these have helped usher in the era of glyphosate's reign. Glyphosate is now king, but like all of its predecessor chemicals at Monsanto, it has many potencies that may be producing different effects in different places. Ambiguity around glyphosate's potencies emerged, I would argue, from its own ontological multiplicity. This multiplicity is another reason for its swirly pervasiveness.

Objects brought into being . . . are *realized* in the course of a certain practical activity, and when that happens, they crystallize, provisionally, a particular reality. . . . They invoke the temporary action of a set of circumstances.
—Steve Woolgar and Daniel Neyland, *Mundane Governance: Ontology and Accountability* (2014)

Ontological Multiplicity & Glyphosate's Safety

From glyphosate's origins as an inorganic pipe cleaner to its rise to glory in genetic partnership as a chemobiological weed-killing agent in agrochemical industrialism, it has been able to travel to many environments and ecosystems. In different places, its potencies have been different. Thus, even though it is but a single chemical, glyphosate has active and animated material lives in different ecosystems of living and nonliving things, from soils to plants to foods to bodies. In each context, glyphosate operates differently according to what it does and what it is put into relation with: eliminating corrosive aging on metal surfaces, turning something nonliving (a chemical) into something that gives life (glyphosate-ready plants) or turning what is living (non-GE plants) into biological corpses, and, possibly, becoming a biologically active microbe-killing agent in human guts. It may also be causing cancer. Glyphosate has the kind of ontological multiplicity that Woolgar and Neyland have described (as in this chapter's epigraph) for scientific objects and their enactments in worlds that humans participate in. The things glyphosate becomes by partnering in different ways with different things, places, and constituencies enable it to form different kinds of material and nonmaterial assemblages, altering both life and the conditions for life.

Glyphosate moves with viral potency, to use Povinelli's (2016) term, from its presence in chemical laboratories to its place as a prolific agent in agroindustrial empires and environments by turning the living into something that is dead in order to let other chemically altered living things live.[1] But glyphosate is powerful partly because it is an interloper across many terrains. It is a chelator here, an herbicide there, a microbe killer here, a carcinogen there. These different potencies become important as glyphosate is conscripted as an interlocutor by different constituencies. Its multiplicity also makes it hard to know, for sure, how toxic it might be across different spheres if all of its potencies are not simultaneously considered. Competing silos of knowledge about its potential harms emerge around it being both a nonliving chemical and a biological partner to many forms of life. These silos have made it unmanageable. In this sense, I would call it not just a viral but also a *vital* interlocutor in the swirl.

We see this multiplicity in action around the question of glyphosate's safety, a question tied inextricably the debates over the safety of GE foods. One way to get at this set of questions is to scrutinize the contradictory relationship between the patentability of these crops and the regulation of them. In chapter 5, I dive into the fraught scientific research on glyphosate and GE foods. Here, I want to focus on how the ontological multiplicity of glyphosate enables it to create swooping recurrences of competing certainty about its potential harm.

Substantial Equivalence

Deliberation over glyphosate's safety to humans (as opposed to environments, microbes, farms, etc.) often devolves to questions of how safe the foods designed to be grown with it are. That question is usually posed thusly: Do foods designed to be grown with glyphosate warrant any more regulatory scrutiny than foods without this characteristic? The answer to this question will be familiar to most readers who have followed the multiple contours of concern over GE foods, but for readers who are new to the topic, I revisit it at some length here because it reveals important insights about how glyphosate's multiplicity helped create controversy about its safety as both chemical and partner to foods.

Most scientific descriptions of GE foods (like the ones I had read about before meeting Michelle) state that glyphosate-ready foods are not different enough from foods grown using selective cultivation, cross-pollination, or other non-GE agricultural techniques to warrant special concern or regulation regarding their safety. Bt crops, in contrast, are pesticides, so they are regulated as pesticides are (NASEM 2016). The claim of substantial equivalence for glyphosate-ready foods has been used in most (though not all) regulatory approaches to them as well as in most of the science on which regulation is based. I am not interested in adjudicating the classification of these foods, but rather in describing how glyphosate's ability to move so seamlessly between the worlds of chemistry and biology enabled it to dodge thorny questions about exactly what it is and how much we should worry about its presence in foods.

Agrochemical foods were born from entangled relationships with agro-capitalism such that even though the patenting of foods was not new and had been done for hybrid crops including fruits and vegetables (Kloppen-berg 2005), the patenting of GE foods *was only possible because these foods were deemed different enough from non-GE foods because of their genetic modifications.* Thus even though glyphosate-rich foods are not considered different enough from other foods to warrant regulation, their difference is precisely what enabled Monsanto (and other companies) to patent and make money from them. The dualism in the definition of these foods—that they are both the same and not the same as non-GE foods—partly revolves around the problem of whether we see these foods as chemicals or as biological things (a debate that was debated in the 1970s as researchers were developing genetic modi-fication technologies for living things).

The precedent-setting case for the patentability of genetically modified organisms was a *Pseudomonas* bacterium created by Paul Berg that could in-gest oil (Kevles 2002). The significance of this case was not just that it was the first time that courts of law in the United States allowed the patenting of life forms. It was also that, because oil-eating bacteria are found in nature, patenting genetically modified organisms that replicate nature in this way meant deciphering hairsplitting distinctions between sameness and differ-ence. Differences that could not be seen at the level of biology could be seen at the level of chemistry. The reasoning used by the judge in this case is telling:

> On October 6, 1977, the Court ruled three to two in favor of Berg. The majority opinion was delivered by Judge Giles S. Rich, who, before his appointment to the federal bench, in 1956, had distinguished himself as a patent attorney during some thirty years of practice in New York City and who manifestly recognized that, to a considerable extent, *life was chemistry.* Rich viewed it as "illogical" to allow patents for processes that relied upon the functions of living organisms but to deny patents to a liv-ing manufacture or new composition of matter as such. *He contended that*

39

in their nature and commercial uses biologically pure cultures of micro-organisms were "much more akin to inanimate chemical compositions such as reactants, reagents, and catalysts than they are to horses and honeybees or raspberries and roses." He found nothing in the language of the patent laws that excluded such tools from patent protection solely on grounds of their being alive; it was being alive that made them useful. He held, "In short, we think the fact that micro-organisms, as distinguished from chemical compounds, are alive is a distinction *without legal significance*." (Kevles 2002, 23; emphasis added)

This patent case broke open the dam that was holding back a flood of patents for genetically modified organisms and gave rise to a whole new platform for public-private partnerships and collaborations of industry with academe in what Fox-Keller (2009) has called "the century of the gene" (see also Landecker 2010). The case was settled in favor of the chemical company's interests by focusing on the key ways a living thing could be seen as a chemical (that is, its DNA was a chemical array) that the company could patent (Moschini 2010), reproducing the message Monsanto tried to convey in its public relations campaigns.

Genetically engineered foods became double at their very inception: both *just like* other foods genetically and biologically and *entirely different* from them chemically. One could say, following Sarah Franklin (2007), that they were *conditional* biological entities such that their genetic alterations became the condition of their profitability and reproducibility. But this condition was erased in the world of regulation. What was, in the world of patent rights, a chemical innovation was also, in the world of regulatory concern, a non-novel biological reproduction. Monsanto was able to work with and profit from GE foods' status as patentable chemical plants (thus holding a monopoly on their reproduction) without being held accountable for how they would behave as biologically active agents that were, for regulatory purposes, portrayed as substantially equivalent to and thus riskier than any other foods.

Keep in mind that if the regulation of GE foods had been based on their action as chemicals that would be eaten by humans, they would have been subjected to much more careful scrutiny and formal clinical trials in humans, as pharmaceuticals are (Perro and Adams 2017). This includes their intrinsic chemical alterations *and* their ability to be grown with and absorb new chemical partners.

Capitalizing on different ways that scientific objects can be purposed or repurposed has always been part of scientific capitalism. It exemplifies what Michael Callon and others call *singularization* in and through distinctions, which work to create profits, as opposed to standardization, which makes things commensurable (Callon et al. 2002). But, in this case, the singularization of glyphosate-tolerant foods enabled the patenting of them even while similar claims of standardization (portraying them as commensurate with non-GE foods) were used to mute concerns about their safety. The difference between food as a unique and patentable chemical and food as an equivalent and safe biological thing is as much about what food *does* as about what food *is*.

Ontological multiplicity (in relation to biology and chemistry) of both glyphosate and the foods designed to be grown with it has enabled a multiplying of their qualities of sameness and difference in relation to other foods and many other chemicals in many different contexts. Multiplicity is a product of their creation story. The chemical glyphosate enabled certain possibilities by being able to be seen as a chemical transaction within plants designed to absorb it, and it enabled companies to convince regulators that, as a biologically partnered part of plants, its ability to jump across different kinds of chemical and biological boundaries made it no different from any other food. This duality of the glyphosate-rich food assemblage has blurred its reputation from the get-go.

Arcuri and Hendlin (2019) point to this problem when they argue that laboratory sciences that reify anthropocentric models of chemical effects overlook how potent chemicals can be in real biological ecosystems. Their case in point is glyphosate. Researchers have shown that the results of glyphosate's effects in laboratory environments are different from those produced in studies of animals affected by glyphosate in the wild. Since the invention of Roundup Ready crops, generations of environmental scientists have been trying to assess glyphosate's presence and impacts in various environments and on various species where it is used (Schuette 1998; Relyea 2005; EFSA 2015), and multiple activist constituencies have formed around these impacts.

For instance, to protect its intellectual property, Monsanto pays what critics call "gene police" (or "seed police") to ensure that farmers are not reusing their patented seeds, and farmers whose non-GE crops have been contaminated by GE strains are subjected to lawsuits for patent infringement (Weiss 1999; Shiva 2000). Patent infringement lawsuits continue to occur globally and have put more than a few non-GE farmers out of business in

41

the United States and worse in other countries (Dowd-Uribe 2014; Stone 2002; Lo 2013).

Then there is the problem of weed resistance. As both weeds and insect pests have developed resistance to glyphosate and Bt toxins encoded in crop plant DNA (an effect of their biological potency), farmers have been forced to rely on more aggressive amounts, more potent pesticides, and newer lines of genetically modified seeds each year. Farmers' dependencies on agrochemical companies for seeds, insecticides, and herbicides continue to escalate (Jeschke n.d.; Kloppenberg 2005). As Birgit Muller notes, "Glyphosate is like an entry drug to chemical dependency" for farmers (Muller 2020, 17). Because of glyphosate's pervasive use, farmers around the world are now in an arms race against weeds (Gould 2018). In 2018, Monsanto recommended, along with the use of glyphosate-based herbicides, new formulations of stronger and more potent herbicides, such as 2,4-D (returning to one of the ingredients in Agent Orange) and, again, a closely related organochloride (known to be neurotoxic) called *dicamba* (Muller 2020). Seeds that are genetically engineered to be dicamba resistant, often in conjunction with other traits such as Roundup resistance and Bt, are already in commercial use (Jeschke n.d.).

This is all before we get to the problem of pesticide-induced soil nutrient depletion and the need for costly fertilizers that arises therefrom. Andrew Kimbrell, from the Center for Food Safety, calls the GE model a "zombie paradigm" for growing food (personal communication; see also Kannisery et al. 2019). By this one could imagine he means not just that GE seeds and their pesticides first kill and then must revivify the microbial life of the soil, but also that the triple dependency of farmers on agrochemical seeds, pesticides, and fertilizers turns farmers into a zombie-like servile class for chemical companies with little say about how they grow their crops and at what cost (though this image would be contested by many agroindustrial farmers). The zombie paradigm could also suggest that companies themselves are like zombies whose life is sustained by the death of ecosystems and the profits this death enables.

One could go down many avenues of critique here. Vandana Shiva describes the GE food revolution as devastating, a "Second Green Revolution," that has nothing to do with the Monsanto claim that GE foods will enable us to feed the increasing world population (Shiva 2000). She has famously said, in Marie-Monique Robin's 2008 documentary film *The World According to Monsanto*, "If they control seed, they control food. They know it. It is strategic. It is more powerful than bombs, more powerful than guns. This is the best way to control the populations of the world." GE food agriculture comes with chemical risks that are seldom tabulated in the cost-benefit analyses of farmers or national policy makers. Lindsay Ofrias (2017) argues that agrochemical industries deliberately orchestrate their research and investment priorities in ways that incentivize contamination from harmful chemicals even while evading responsibility for these contaminations (see also Harrison 2006). This becomes particularly clear in the offshoring of chemical harm to poor countries and among poor communities where chemical companies' ability to efface recognizability and accountability for harm from their products is most visible, which has been very true for agrochemical products (Lyons 2018; Agard-Jones 2014; Nading 2015; Dowd-Uribe 2014; Stone 2002, 2010; Avila-Vazquez et al. 2017; Sanchez Barbra 2020). The externalizing of responsibility by displacing blame and the use of concepts such as collateral damage all work to allow ongoing chemical exposures (Gilham 2017; McHenry 2018; Klein and Hamerschlag 2016).

I do not want to belabor the litany of complaints against the agrochemical industrial empire, though there is very good research and writing on this topic. My point is that some of this agitation arises from a more subtle point of departure that has to do with the chemicals and the ambiguous nature of these food crops—an ambiguity that arose with their creation and that has permitted them to move between material and nonmaterial, living and nonliving systems in ways that diffuse our ability to know what they are and how they should be calibrated in relation to harm, forming a swirling mess of competing claims and unstable certainties. To understand this in relation to the presence of glyphosate in food, it may help to know a little more about how the multiple ontologies of GE foods and glyphosate enable them to slip through regulatory gaps.

Regulatory Slippage

As a GE food companion, glyphosate circulates in several realms of regulatory scrutiny, including in and through agencies responsible for regulating food, pesticides, chemical harm, and environmental security. Here, glyphosate is simultaneously a pesticide, a food ingredient, an environmental chemical, and a biological agent used in research with laboratory animals. It is in part this multiplicity that has created problems for it. When it is a GE food ingredient, it is regulated one way. When it is a chemical pesticide, it is regulated another way. Its travels as both a food companion and a weed killer have given rise to ample research studies and efforts to translate between them to determine how safe it is in different places—specifically homing in on it as a

43

possible carcinogen—and in that context it is regulated differently still. No studies or agencies explore its capacities as all these things: chelator, bacteria killer, resistance-inducing competitor, soil bacteria depleter.

Regulations of GE foods and glyphosate differ from country to country; even within the United States and Europe they can vary from state to state, county to county, and municipality to municipality. In the United Kingdom and Europe, public and scientific research showing potential dangers of GE foods led to the adoption of a precautionary principle that restricted their entry into the market until more was known about their risks in agriculture and in the commercial sale of foods for humans or animals (Schurman and Munro 2010). To this day, GE seeds are not allowed in some European countries, but not all, and glyphosate is banned in some countries but not all (Arcuri and Hendlin 2019). Even in some regions that have a total ban on the use of glyphosate, large agricultural regions right next door may use it robustly, often causing "drift" into no-glyphosate farming areas (Ackerman-Leist 2017; Baum Hedlund 2022). In the United States, there was no adoption of a precautionary principle. Instead, the push to bring GE foods to consumers was swift, and the ability to restrict them on any grounds at all quickly left the hands of scientists and was turned over to regulatory agencies (Druker 2015). US regulatory agencies, relying almost entirely on what industry scientists told them, did little or nothing to regulate glyphosate-ready foods (Martineau 2001; Muller 2020).

Glyphosate and the foods designed to go with them at times slip through regulatory gaps. The United States provides a good example of how this occurs. Regulatory agencies rely on scientific studies produced by industry, government and nongovernment research institutes, academic institutions, and testing agencies that are largely, as critics have noted, staffed by industry scientists (Bo 2006; Gilham 2017; Oreskes and Conway 2011; Davis 2014). Only some of the industry studies used by these agencies are published and only some are available to the public. This alone leads to confusion about actual safety simply because so many of the studies are not scrutinized by scientists outside of industry or whose institutes or departments have not received industry funding. When industry reports are made available it is usually only after the agencies were petitioned or sued for access under the Freedom of Information Act, often long after the reports have been used in regulatory decision-making. Indeed, most of Monsanto's research papers concerning the safety of glyphosate were only made available to a wider public (including academic researchers) starting in 2017 when the Monsanto papers became available (McHenry 2018).

44

Added to these gaps are those created by the architectures of responsibility in these agencies. Toxic chemicals in foods, if designated as toxic at all, are regulated by the FDA, but the FDA does not consider GE foods to be different enough from any other food to warrant special regulation. In other words, the FDA adopted the industry view that genetically engineered foods were no riskier than foods that were not genetically engineered in terms of their innate composition (that is, the FDA deemed them "substantially equivalent"), and thus should be, like any other food, "generally recognized as safe," or GRAS, and need no special regulation from the FDA on this basis. The FDA does care about how many pesticides show up in food, but it follows the EPA guidelines to determine how much of any pesticide is safe. Bt foods are regulated as pesticides, but not by the FDA.

The USDA (US Department of Agriculture) regulates pesticides used in animal feed, and the BLM (Bureau of Land Management) regulates pesticides in fisheries and wildlife areas. However, both the USDA's and BLM's regulations on glyphosate have been governed by the consensus view that the foods are safe following GRAS guidelines and that glyphosate is not toxic to animals. These agencies do pay attention to studies that identify the degree to which glyphosate bioaccumulates in different environments, adopting the scientific reports that say it does not bioaccumulate in forests, fields, or water, and thus poses little or no risk to animals or humans when used at the current levels recommended by the EPA (NASEM 2017). As a reminder, this is based on the argument that it is soluble in water but, because it binds tightly to soil, will be broken down after up to six months by bacteria present in the soil—in other words, it will act like other chemicals that can biodegrade. Critics argue that, in this time frame, glyphosate can do a great deal of damage to living creatures; but, more importantly, the saturation levels from repeated spraying of glyphosate products within and beyond six-month intervals makes the question of low bioaccumulation moot.

Since the EPA regulates pesticides, it regulates both Bt crops and glyphosate using available scientific studies that have determined safe or allowable levels. The determinations for such levels can mean different things. The reports used by the EPA to determine allowable levels rely on things like "evidence of harm" that tend to be very narrowly construed. In fact, what *evidence of harm* means requires determining not only what *harm* means but how it is determined for different kinds of chemicals in different kinds of animals and environments. As others have noted (Murphy 2006; Boudia 2016), regulatory science is organized around reductionistic models that prefer to isolate one chemical at a time. If a chemical is shown to cause harm

45

at some level, regulations are set to limit its use beyond those levels and to post warning labels about the possible harms it is known to cause at amounts above those levels.

Even when regulatory agencies do their best to set policy to protect the public, wildlife, and the environment from chemical harm, we might still be concerned about the gaps that emerge in the infrastructures of knowledge based on the statistical technologies used to decipher harm for chemicals versus living things. Consider the complex algorithms used in different kinds of regulatory decision-making today. To determine allowable levels of chemicals used in the food production system, the EPA relies on something called no significant risk level (NSRL). Just what "no significant risk level" means is complicated and often requires extrapolation and estimation. The EPA's regulation of glyphosate relies on different measures depending on whether one is looking at the science that talks about pesticide levels in the environment, field crops, foods, laboratory animals, or human bodies. And even though the EPA produces guidelines and establishes policies about safety, the practices of determining safety levels about food can also vary in the United States from state to state. In some studies, toxic effects are characterized in terms of NOAELs (no observable adverse effect levels)—the maximum amount of a substance that can be delivered to a test subject (in this case a rat) without adverse effects appearing.[2] Again, how this level is ascertained is not standardized and can differ from state to state and from one kind of observable effect to another. The California regional EPA office, where I observed deliberations, for instance, uses safe harbor levels, which include NSRLs for cancer-causing chemicals and MADLs (maximum allowable dose levels) for chemicals causing reproductive toxicity. These levels are based on observations of increased risk of developing cancer as a percentage of the normal rate for tumorigenesis in laboratory animals. For the risk of health effects from chronic chemical exposure other than cancer and gene mutations, the EPA uses chronic reference doses (cRfD). This concept replaced, in order to consolidate, previous measures such as acceptable daily intake, safety factor, and margin of safety. The EPA also works with *tolerance*, the maximum residue limit, which is the amount of pesticide residue allowed to remain in or on each treated food commodity (which is determined based on studies that show MADLs and NSRLs, among other things). Finally, researchers use parts per billion (ppb) to ascertain how much of a chemical in terms of parts per billion is in the food and drink people consume based on the levels present in crops themselves. Although some researchers are trying to look at these levels in consumable food items, the laboratory techniques for determining these are also imperfect and run a high risk of false positives. Still, there is a conversion for ppb in crops to milligrams per kilo per day in human consumption, so this measure is considered useful to regulators—if they believe the studies that show a chemical is harmful at all, which is often not the case.

The complex translations between evidence from laboratory animals, crop composition, consumable foods, and humans to determine safety, in addition to the need for keeping track of the conversions between the acronyms in play in the question of allowable levels, create a good deal of obfuscation and some uncertainty even for regulators. Despite the hairsplitting precision offered in the statistically built technologies, it can still be hard to determine actual risk because it becomes difficult to know which measures to use and how to set them in conversation with other measures. The science is certainly not something the nonexpert can grasp without help. This is why EPA panels consist of experts who do know how to read these reports. Still, even these experts often rely on the reports' conclusions and summaries, reminding us of the importance of who is writing and producing them. Using these reductionistic and complex techniques, and despite a growing archive of concerns over glyphosate's toxicity (as we will see in the next chapters), the EPA concluded in 2020 that glyphosate was safe: "After a thorough review of the best available science, as required under the Federal Insecticide, Fungicide, and Rodenticide Act, the EPA has concluded that there are no risks of concern to human health when glyphosate is used according to the label and that it is not a carcinogen" (EPA 2020, 1).

This brings us back to the problem of regulatory capture—that is, that regulatory agencies have revolving doors for chemical industry experts who cycle between companies and regulatory agencies and "capture" the regulatory process so that it works to benefit company interests (Oreskes and Conway 2011; Boudia and Jas 2014a; Gilham 2017). Jonathan Latham, head of the Biosciences Resource Project in London, argues that revolving-door relationships between companies and testing agencies may compromise companies' ability to be honest, but this process alone cannot encompass the complex ways that regulatory agencies do what companies want. Using the case of dioxin, he argues that the main problem with the regulatory system is not a process of capture but of proactive collusion. Chemical companies create hired-gun testing companies that produce scientific reports in order to create the appearance of disinterested science and to give the illusion of a firewall between the scientists at the testing companies and regulators. Nevertheless, the fact that the testing companies are propped up and paid for

47

by industry is not disclosed to regulatory agencies and does not need to be. Thus, regulatory agencies collude with industry when they fail to scrutinize the quality and findings of the testing companies' research, even when it is quite flawed, and instead take these companies' summaries and conclusions as factual and reliable.[3]

Even if we had better regulation, as Boudia and Jas (2014a) have argued, more regulation (of any given chemical) does not necessarily mean more protection against harmful chemicals. Relying on scientific studies to decipher harm from toxic chemicals is challenging for all constituencies, not just regulators, because the scientific archives often produce a kind of ignorance. And regulation is seldom simply a matter of experts relying on the available science. Regulatory agencies also must contend with scholars and activists who have been pushing against the use of scientific peer-reviewed publications as the only arbiter of the facts about chemical harm (Krimsky 2014; Frickel and Moore 2015).

One could produce ample evidence to argue that regulatory capture and proactive collusion are at work in producing and spreading glyphosate to unprecedented levels in the United States. To be sure, many blind eyes have turned away from potential sites of toxicity because the commitments of industry scientists pushed some agendas and produced some kinds of data but not other kinds. Regulators may even be producing and relying on flawed studies that have been bent to industry's goals. But my point is not this. My point is that the offices of the EPA (at both federal and state levels) rely on practices of determining safe levels of toxic chemicals that reproduce the same diffracted patterns of evidence that go into making and justifying the use of toxic chemicals in the first place (Davis 2014). This partly has to do with regulatory agencies' myopic focus on isolated chemicals and specific effects—mostly carcinogenicity (Murphy 2006; Boudia and Jas 2014b).

In the case of glyphosate-rich foods, the problem is more complicated. The statistical and technological gymnastics that go into producing studies for the EPA are sophisticated, and one might say they offer the best tools we have to decipher the safety of chemicals. But, despite the hairsplitting sophistication of these technologies, agencies like the EPA must still rely on a process of interpretation to decide how much to regulate any given chemical, and these agencies cannot treat pesticide-rich foods as a problem of biologically active agents instead of as a problem of chemicals. In this sense, glyphosate's intrinsic multiplicity enables it to escape much regulation because its potencies are diffused into and diffracted onto different kinds of material things and places. Glyphosate is seen as a nonissue by the EPA because the studies they rely on focus mostly on its travels as a chemical and not as a living thing or part of an ecosystem that takes it up and uses it in ways that keep it "alive." Since glyphosate actually arrives in the human body with food, one might say that it is part of a metabolic system that the FDA should care about, specifically in relation to its ability to do things other than kill weeds (traceable through metabolomics and assays that might show its potencies as a microbe-killing agent). Even if one believes it is noncarcinogenic in foods at specific levels, what might these noncarcinogenic levels be doing beyond not causing cancer? Studies on the metabolic potency of glyphosate in food are not used by the FDA because the foods glyphosate is designed to be used on are not considered different from non-glyphosate-rich foods and because the FDA doesn't study these things for any foods. Rather, it is concerned only with chemicals that come in foods as chemicals, sui generis, and the EPA has decided that glyphosate—as a chemical—is safe when used properly. But glyphosate is also a lively agent in food ecosystems.

Glyphosate lives in soil in ways that other chemicals do, eventually being broken down by bacteria, but the possibility that it is harming soil bacteria while biodegrading, and the idea that the minerals it binds to are being taken away from food plants along the way, too, are not of concern for regulators because this has been happening since the dawn of agrochemical industrialism (and because there are remedies to it: the industries that make soil-depleting agents also sell farmers soil-enriching agents). If glyphosate were seen as a lively provocateur (or even a terrorist) that spreads itself into many places in ways that augment some life and eliminate others, how might regulations change? If the potency of glyphosate were characterized as alive—a biological agent of great potential—might this force regulatory studies to focus on ecosystem effects that can be seen all the way up the food chain, as Arcuri and Hendlin (2019, 2020) suggest?

I am arguing here that regulation is stymied by forms of chemical reductionism that come with a lot of fancy but ultimately obscuring statistical footwork. But I am also arguing that glyphosate, by matter of its existential relationship to food systems, is more than a single chemical and more than a chemical. It is a living thing that partners with other living things in labs, soils, foods, and bodies, and because of its molecular structure works as a chemical that is able to effect changes on other living and nonliving things. Glyphosate falls through the gaps of regulation in terms of its biological impacts even while being carefully regulated as a chemical, and falls through

49

gaps in chemical regulation by being treated as biologically equivalent to other living things when in plants. Its multiplicity makes it an unstable and unmanageable thing.

In the end, we might say that glyphosate moves back and forth between the worlds of biological and chemical activity and shape-shifts along the way such that anyone or any institution interested in tracking it can, without much friction, make it empirically productive in the ways that those constituents need it to be. This multiplicity at times leads to regulatory omission and effacing and other times to regulatory practices that enable one or another of its potencies to prevail. The multiplicity of glyphosate has been one of its most ardent advocates, creating swirls of contradiction and concern in its wake but seldom creating singular affirmations of its potential harm or need for stricter management.

This same problem arises in debates about glyphosate's safety for humans, giving rise to an archive and a set of practices that form new resting points in the swirl. Before turning to how the scientific archives on glyphosate form a swirl, I want to return to the ways that biological effects of glyphosate show up in human places where its chemical potencies have begun to spark debate: clinical care. In these places, this little chemical creates new sensoria and sensibilities—swirls in their own right—that are forged in the relationships it creates between clinicians, patients, and therapeutic hope.

I'm the product of my mother and father's DNA and of the DDE [dichloro-diphenyldichloroethylene; similar to DDT]. These lay down with them in the nuptial chamber. A mouthful of breastmilk and DDE formed my first human meal. At the resurrection of the body we will carry these along with our sins to the throne of judgment. Peter will weigh them: sins, graces, pesticides, radiation. —Susannah Antonetta, *Body Toxic* (2001)

Chemical Life, Clinical Encounters

In her pseudonymous memoir, *Body Toxic*, Susannah Antonetta reminds us that the sequestering of chemical evidence in and through bodies hides itself in surreptitious ways, sometimes arriving before birth. Chemicals ingested with food are particularly insidious in their capacity for concealing the origins of their harm. Toxic ingredients are usually not visible to those eating them, and harm from them can accumulate and cluster in bodily tissues long after being ingested but before symptoms indicating that things are going awry present themselves. Intimate life with chemicals works this way, with submissions and absorptions, interferences in complex physiology and a prompting of functional decline remaining just out of conscious reach, until their effects cannot be silenced, the evidence popping up on the surfaces of skin or in digestive irregularities, persistent brain fog, cancer. Chemical effects invite us to become fluent in languages of probabilities and possibilities and to learn to hear their presence through repertoires of chronicity. Glyphosate offers a similar story. It is hard to know for sure what the effects are of small doses of glyphosate absorbed in human bodies through gardens tended or foods eaten not once but frequently over many years, but one thing is likely: they will appear in chronic form.

Chronic ailments now become us, Lochlann Jain (2013) notes for cancer, churning in our cells as slow-to-grow mutations alongside the breaks and sprains, infections, flus, and colds. Chronic ailments from chemicals are a source of frustration for patients and clinicians because they hide their truths and their origins. They surprise, suddenly appearing as new ailments even though they have been brewing a long time, lurking in tissues that we didn't know were harmed, possibly from one chemical or another or by something else. Chemicals creep in and stir harms that are at first mostly tolerable and manageable and often don't have names that are specific enough to warrant a diagnosis or treatment. Sometimes symptoms can be quelled, even if the harm persists, making treatments feel like cures—until they stop working. Rashes, recurring diarrhea, the creeping severity of asthma, brain fog that gradually worsens for no apparent reason, autoimmunities, mysterious lumps . . . These are the problems one learns to live with, the initially inconvenient and then suddenly taking-pills-every-day-for-them problems; or worse, the problems for which surgery is required, or for which there is no treatment, let alone a name, until it is too late. Then they take on an urgent form when they are given a name: cancer, colitis, depression, stroke. But the harm has been there in the body all along, only sometimes giving rise to a sense perception of harm. As Nicholas Shapiro reminds us, "It is by virtue of this very capacity to be chemically wounded, even minutely so, that bodies bear revelatory power" (2015, 370). The question is, he says, how do we attune ourselves to sensing this harm? What sensibilities and forms of reason might chemical harm demand and how might this sensibility become operationalized in clinical settings? The archive of social science thinking about chemical harm is helpful.

Chemical harm from glyphosate is hard to trace in part because it is hard to decipher harm from any chemical at all. Sense perceptions can be misled by the absence of concrete scientific evidence and the inability of clinicians to suspect chemical harm in anything but high-exposure cases that present as acute. The problem goes far beyond chemical harm from agrochemical industrial food. Toxicants in the form of PCBs, phthalates, parathion, parabens, pesticides, PFOS and PFOA (to name only a few that begin with the letter p) and uncountable other chemicals now saturate water supplies, air, clothing, furniture, electronics, and buildings (Gaber 2019). The ubiquity of exposure is noted abundantly in the literature, with the accumulation of low-dose chemical exposures from many sources moving us collectively from "the silent spring to the silent night" (Hayes et al. 2002; Hayes and Hanson 2017) and to a new age of toxicity (Walker 2011; Liboiron et al. 2017), with a multitude of suspected and known distortions of bodies, animals, and living environments. One's sense of purpose around chemical exposures can become easily derailed by a persistent inability to gain traction on the reasons they have been allowed to persist, from regulatory to clinical obscuration. Given the

52

intractability, scholars have recently been thinking about chemical harm in relation to questions about ethical research and a need to refuse dreams of chemical-free purity (Liboiron et al. 2017; Balayannis and Garnett 2020; Chen 2011). When scientific and laboratory forms of evidence are unavailable, we might consider community-based research in relation to designs, responses, and priorities such that knowledge production is aligned with activist goals, treating chemical pollution as a form of colonialism (Liboiron et al. 2017). To acknowledge the uneven distribution of chemical harm along lines of race and class means recognizing that evidentiary platforms for action against chemical injury must also sometimes accommodate alternative ways of knowing and experiencing injury (Shapiro 2015; Agard-Jones 2014; Hoover 2017; Shadaan and Murphy 2020).

Doing ethical research on chemical harm can mean moving beyond assumptions that there are innocent victims and guilty perpetrators in the chemosocialities we now inhabit (Ticktin 2017). This is true not just because many of our most toxic chemicals are also helpful to our chosen ways of life, but also because our very efforts to eradicate them can reproduce the infrastructures that enable them to persist, even when we know they are harmful. Most people participate in regimes of chemical exposure because of Enlightenment-era attachments to chemical potencies that are shared even by those who are harmed by them (Masco 2006). Moving beyond such attachments includes, among other things, avoiding utopian dreams of chemical-free purity and recognizing that the institutional commitments of many activisms have already mostly failed (Taylor 2017; Chen 2011).

Engaging in an effort to eradicate harmful chemicals might thus require working along several different lines of engagement: first, one that requires a rethinking of the way we tell histories, including the roles of settler colonialism, capitalism, and racism and their fictions of security (Murphy 2017b, 2018b Shadaan and Murphy 2020; Liboiron 2021); and, second, one that reconfigures noninnocence (and nonpathologizing) of the harm that is already done while still pursuing nodes of engagement that point to sites for repair and accountability. One of the most urgent commitments in a post-Enlightenment approach to chemical harm is perhaps the effort to move past the problematic adherence to the idea that even if science got us into this mess, it can still get us out of it. The push/pull of science among those trying to gain accountability and redress in relation to chemical harm is a recurring theme in this book, and my goal is to read this as partly an artifact of the chemical potencies of glyphosate—turning to the history of the chemical as forming a swirl of arbitrations among multiple constituencies that are also

53

governed in part by the chemical itself. All of these concerns are magnified in their ability to confound concerns about both knowledge and accountability when it comes to chemicals that arrive with food.

Chemical exposure from food holds a special place in the lineup of elusive toxicities. Toxicity from food is bundled in uncertainty because of the way evidence of harm from food is doubly fraught by the politics of evasion and displacement tied to the industries that make it and the circumstances that produced it (Guthman 2019; Davis 2014; Creager and Gaudillière 2021; Stone 2010; Saxton 2015; Kloppenberg 2005; Gemmill et al. 2014; MacLeish and Wool 2018). To be sure, food is not just a consumer good; it is a vital infrastructure that has, since at least the birth of industrial agriculture, invited scrutiny over how its safety is ensured by way of an acute set of discernments that are entangled with problems of capitalist corruption and evasions (Guthman 2011, 2014; Nestle 2018; Kessler 2010). And, as I said, the harms of agroindustrial chemicals are not distributed evenly; they have intersectional effects that fall along lines of race, class, ethnicity, and gender, and these uneven distributions have often impeded a sense of urgency about them in a very uneven politics of expendability (Eskenazi et al. 1999; Saxton 2015; Hoover 2017; Sanchez Barbra 2020; Hendlin 2021). The harm swirls and lands unevenly in environments, targeting weeds but being absorbed by edible crops and coming into different bodies even within single households in uneven amounts. Because they are infrastructural, foods also spread their potential toxicities widely and up lines of privilege, even when food insecurities and inequities may exacerbate the problems of chemical exposure. Where they go, they can wreak havoc, but in ways that are not uniform and not reducible to problems of gender, race, or class. These movements confound those who have gotten used to thinking they are immune to chemical toxicity by matter of their privilege but who now find themselves trying to sleuth, heal, and help themselves or members of their families as they wade through oceans of possibility in sourcing the bodily damage that may be arising from their food.

Deciphering toxicity from food is challenging because of the way chemical harm that comes with foods works its way into and through bodies (Sutton et al. 2011; Sherman 1996). Brett Walker writes that chemical toxicants like insecticides penetrate whole ecosystems and thus "have the potential for more deep-seated alterations of the environment and human bodies" (2010, 55). But, because of this saturation, the root causes of these alterations are often masked by their pervasive and varied appearances across ecosystems outside and inside of bodies. Chemical harm from food is notoriously hard to track in human bodies in specific ways, especially ones that link one chemical to more than one kind of damage (Murphy 2006; Chen 2011; Kenner 2018; Shapiro 2015; Lamoreaux 2016). Chemicals often change form after being digested and absorbed, sometimes becoming undetectable after a period of time even though they have altered the cells they have touched and the molecules they have bonded with and/or mutated. Even when chemicals cannot be detected in things like soils or the body's waste, this does not mean they have not left other traces behind or altered what they touched as they moved from dirt to microbe, gut to blood to skin, brain, kidneys, livers, lymph. Still, different appearances of harm can lead to multiple suspicions of causality, with pesticide chemicals being only one possibility.

It is hard to know for sure that any bodily damage is caused by one chemical in some foods, or many, or even simply by the food itself regardless of its chemical content. Like environmental pollutants, chemicals in the food system remain on the move and are often unstable interlocutors for activism or scientists (Liboiron et al. 2017), leaving behind heterogenous traces that work to displace and relocate harm (Balayannis 2020). Regulatory agencies tend to reduce concern to singular chemicals, but as Hendlin (2021) notes, following Jasanoff's technosocial imaginaries, "dominating chemical imaginaries compartmentalize chemical elements for regulation, failing to attend to the dysergistic cocktail effects of actual chemical compound exposures and bioaccumulation" (182). This is before we get to the fact that many a scientific consensus (not necessarily industry funded) says some of these chemicals arriving with food are not toxic at all, at least not in amounts prescribed as safe.

The effects of food-related chemicals are many and varied since food itself is designed to be used in many and varied ways in the body. Food is metabolized into different vital materialities and functionalities. Endocrine modulation, intestinal absorption, amino acid production, synaptic firings, the viscosity of mucus that softens the movement of joints, the ability of alveoli to pass oxygen to the bloodstream or generate mitochondrial explosions that pump the heart or allow the immune system to rally white blood cells and rid the body of things that don't belong . . . all these entail the transformation of food into molecules that shift utility as they move to or through different locations in the body. Given that nearly every cell in the body is made up of (or impacted by) the organic and inorganic matter that is ingested, the entirety of the body and all of its physiological systems are potential sites for chemical harm. Thus, chemicals in foods become digestive interlopers in ways that matter more, and in more varied ways, than other chemicals one

55

might be exposed to. They are, in the sense that Hannah Landecker identifies arsenic, metabolically potent (Landecker 2011, 2019; see also Mol 2008), becoming another kind of vital infrastructure inside bodies. This also means that the content and form of bodily things (tissues, cells, metabolites) can, like chemicals, change and be changed by the process of metabolism itself.

The challenge one faces in tracing chemical harm from food is not simply in moving beyond the problem of industry corruption and the diffracting of certainty by an overproduction of competing claims about the evidence. The challenge is in how food-companioned chemicals live in and through bodies, food, and environments. We might think of the heterogeneity of the function and form of food chemicals in bodies as giving rise to new sites for environmental chemical opportunism, forcing new trade-offs between harm and profit, and in the effort to trace them and decipher lines of actionable intervention, forcing choices between injury and benefit. Like the stories of other chemicals, that of glyphosate offers a window into the world of shifting certainties around chemical harm and shines another beacon on the need for new ways of thinking about the facts, politics, and experience of harm in relation to it.

Innovative scholars point to new ways of tracing glyphosate's effects. To somatically apprehend the effects of glyphosate might require, again following Nicholas Shapiro's insights about formaldehyde, attunement to the multiple sensibilities of harm felt in the body's sensorium: rashes in one person, headaches and brain fog in another (Shapiro 2015). Attuning to the chemosphere, as Shapiro calls it, is a task not just for people with exposures; it is also for the clinicians who try to help them, as I will show. Similarly, following Mel Chen's (2011) "toxic methodology," our life with glyphosate enables an iterative engagement with chemicals that promote new ways of fact- and theory-making. By displacing chemical exposure onto a wide range of processes, not just human social relations but also the chemicals we are intimate with, we might consider dispersing notions of pathology onto an ontological field where borders are replaced by interactions, interabsorptions, and mixings, sensibilities of an ever-spiraling set of symptoms and signs. Experiential uncertainty about chemical harm becomes settled in bodies in ways that then become unsettled. Like the facts of chemical harm, they are always on the move. How, then, might that unsettledness inform how we think about chemicals and the infrastructures of care they bring in ways that exceed our current political engagements with them (Murphy 2018c; Balayannis and Garnett 2020; Shapiro and Kirksey 2017)? Glyphosate offers a peculiarly rich opportunity to explore these things.

56

My colleague and coauthor Michelle (the integrative physician) daily confronted patients in her pediatric clinic whom she believed were injured by chemicals like glyphosate. Again, I turn to her cases with caution. My goal is not to privilege the idea that children provoke a special concern about the future, nor am I invested in the moral panic Michelle had over a generational effect—a reproductive futurity in which gendered labors of care by mothers and clinicians would be reproduced. I am interested in how her encounters offer insight into how one might wrestle with the effects of chemical exposures from food as a collaborative endeavor that loops parents, patients, and clinicians into a world where the science and evidence bases are thin but the stakes are high. My hope is that readers will be able to read the stories about children here as sentinel in broader ways—as indicative of the effects that unregulated chemicals have not just on food but on sensibilities about the body, health, and the fraught scientific archive.

Clinical Sensoria

Michelle's clinic was filled with parental and childhood trauma. She typically saw parents and kids who were at the end of their rope. The children had chronic problems that, despite persistent efforts, would not resolve. They had eczema over 60 percent of their bodies, asthma so bad they used inhalers every day, chronic headaches, and recurring diarrhea. They had been diagnosed with reflux, IBS and ulcerative colitis, Crohn's, and some with celiac disease. She saw kids who had sensitivities and allergies to so many foods that nutrition was an everyday uphill battle. This was in addition to the weekly parade of children with diagnoses of neurocognitive problems: some with ADHD and neurodevelopmental delays; young teens with depression, anxiety, and bipolar disorder; and, of course, an overwhelming number of very young children diagnosed with autism spectrum disorder. Parental anguish over dashed dreams of perfectly healthy children loomed large in Michelle's clinic, fueling a sea of outrage and frustration over the adjustments, the pharmaceutical or surgical burdens, and, for some, the unending search for doctors who could give them answers. They wanted to know what caused this harm and how it might be remedied.

The families Michelle worked with had seen multiple doctors and specialists by the time they came to her. They had been given steroids, antihistamines, and analgesics but, despite all the doctor visits, all the school days missed and birthday parties canceled, parents believed they had received only symptom management with no real explanations of the causes or anything close to cures for the failing health of their children. In some cases, parents were told there were no treatments, or that their children's symptoms were not serious despite the truckloads of steroids they had to smear on them every day, despite the daily panic of making sure their inhalers were working and stuffed into their school backpacks, despite being told that they might at some point need surgery for their ulcerated guts. "Kids who were not even twelve years old!" Michelle exclaimed. "Can you imagine?" Michelle would roll her eyes in disbelief, sometimes over the fact that *so many* kids were presenting with these problems, other times because she believed her own medical colleagues had failed these children, had too quickly given up on finding cures and instead prescribed nothing more than symptom-suppressing drugs in what she disparagingly called "pill for ill" medicine. Toxic chemicals in GE foods were simply not on the radar for most of her colleagues.

Michelle's convictions about glyphosate being an underlying cause of these ailments were long in the making. Her journeys into integrative medicine and to the belief that pesticides in foods were a major pathogen began some twenty years earlier when, as an emergency room pediatrician in a local hospital and the mother of two young children, she was asked by other moms in her community to provide medical expertise to their campaign against the neighborhood spraying of a pesticide. She wasn't an expert on it, but she went exploring and came across the work of a Hungarian nutritional biochemist named Arpad Pusztai and began to think about the possible connections between GE foods, pesticides, and the symptoms she was seeing in her patients. Pusztai's work had convinced her that genetically engineered foods, because they had pesticidal effects, were causing inflammation of the gut, dysbiosis, and the array of chronic ailments she was seeing. Michelle believed these children were exposed to toxic chemicals from food that had been genetically modified and now contained glyphosate, among other things.

As I watched Michelle care for her patients and listened to their parents, I gathered stories that would form another chapter in the life of this chemical, about how humans may or may not be coping with glyphosate's many potencies and giving flight to new formations of certainty by conjuring pathological and etiological theories that could explain its role as a root cause in many disorders. One might say that Michelle's work formed a swirl of clinical experience and epistemology, an observation that invites us to follow the swirl at work in bodies of this chemical on the move. The effort to sense glyphosate in bodies means, in other words, reading ailments as heterogenous indicators of the multiple potencies of this chemical. Each patient was apprehended as a palimpsest of signs and symptoms that pointed to the telltale evidence of glyphosate-rich foods, among other things.

Take the story of Melissa and her daughters. Melissa arrived at Michelle's office with her two children: one child in a stroller and a four-year-old named Amelia who walked in wedged between the stroller and her mom, trying to push the stroller herself. Once in the office, Amelia was instantly distracted by the toys in the corner and went right to them. Without skipping a beat, Michelle went to Amelia and offered a big smile, saying, "Hello, Amelia!" Amelia looked up for a moment, smiled and then looked away, turning her attention back to the colorful blocks. Michelle introduced me and Melissa began to tell us both about the remarkable improvement she had seen in Amelia's health. "You wouldn't believe it," she said, turning to me directly. "A few months ago, Amelia was totally miserable. She had chronic constipation, was always bloated, and she complained all the time. She was in pain. She was lethargic, never seemed alert or able to respond even to things I knew would light her up." By now Amelia had crawled up to the examination table and Michelle was taking her vitals. "She seems great, now," I said, observing her active engagement with the doctor and the toys, full of smiles. Her mom continued, "It took a few weeks, but we started seeing improvement right away with her stools being more regular, and you know, not sticky and dark. Her bloating is gone. Her mood is bright. She's like a whole new child."

"I see this a lot," Michelle told me after Melissa and her kids left. "Clean up the gut and get rid of the foods that are triggering immune reactions, stop the inflammation, and the kids get better. It's not brain surgery." It wasn't brain surgery, but neither was it in any way typical of mainstream medicine. Her clinical approach started with a two-hour intake appointment in which she inventoried practically everything she could about her patient's medical history, exposure to toxicants, previous treatments, living environment, the ebbs and flows of different symptoms, and, most important, their typical diet. She asked parents to start a diary of every food their child ate and to track any symptoms they experienced, all with times and dates. She sent parents and kids to the lab for blood tests, urine tests, and food sensitivity and allergy tests. To get the full picture of what bacteria were in the gut, Michelle sometimes requested a genetic test that reported the presence or absence of major bacterial groups appearing in a patient's stool. If need be, she asked for blood and urine toxicology screens to see if there were obvious chemicals playing a role in pathology.

59

Among the most important of the lab tests Michelle ordered for her patients was one that detected food sensitivities. Food sensitivities could be detected through reports that showed immune system responses to specific food prompts. The immune reactions she was looking for were not anaphylactic responses (IgE) but the lesser known responses (IgA or IgG). In her view, the presence of these immunoglobins meant the immune system was being triggered, though many of her mainstream colleagues disagreed with her. Conventional clinical guidelines suggested that these lesser immune responses were likely a normal part of the digestive process, not anything to be worried about. But for Michelle, they pointed to persistent, low-grade chronic inflammation. The chronic provocation of the immune system from chemicals in the blood or the presence of chemicals that killed important gut bacteria led to dysbiosis (an imbalance of undesirable versus desirable gut microbes) and impaired digestion and nutrition, leading to a cascading set of problems.

Even before she read the results of all the tests, Michelle almost always prescribed some key interventions: remove inflammatory foods like sugar, gluten, and dairy, and shift to an all-organic diet to avoid extra chemicals in food. These key interventions, she argued, were the first line of defense: get rid of trigger foods and especially the GE foods, the glyphosate. Sometimes the results of even these first-order interventions were dramatic, even if it was a struggle for families to implement them. Other times these preliminary efforts were just the start of a long journey of hit-and-miss attempts to home in on things that would work. There was no test for glyphosate that she could order, but if other chemicals showed up in the lab results, she worked with parents to figure out their sources—not just from foods but also from environmental exposures at schools, farms, homes, and neighborhood parks. Did they notice spraying of weeds near or on the premises of the day care center? Had they noticed neighbors spraying in their gardens? Did they live in a high-pesticide farmland area? Michelle checked patients for deficiencies of vitamins and minerals and recommended supplements. For treatments, in addition to using mainstream medical pharmaceuticals, albeit sparingly, she also prescribed homeopathic remedies. Sometimes she recommended osteopathic and chiropractic or other manipulative interventions that worked well for children.

60   Using these diagnostic and interventional techniques, Michelle saw headaches subside in young kids; muscle tone and cognitive alertness pick up in babies; rashes disappear, along with constipation, reflux, diarrhea, bloating. Most of her patients were able to go back to school, to have playdates, maybe

to discontinue an inhaler or steroids. Parents were soothed by their child sleeping through the night or getting relief from regular bowel movements instead of constipation (or the opposite, recurring diarrhea). They reported clarity of speech and behavior and no more itching. Parents even talked about improvement in the behavior of children who had been diagnosed with autism spectrum disorder.

For Michelle, food was the base code—the connecting architecture—that linked all of the disorders she saw. What mattered most about the food was whether it was genetically engineered, glyphosate-rich food. She said that what she practiced was "food-focused medicine"—a type of medicine that paid attention to the inherent quality of foods eaten as well as the dietary health of patients. Because her theories of food-based pathology were unusual, she often found she had to educate her patients and their parents about them.

Irene had this experience when she took her son, Sean, to see Michelle. Sean was a three-year-old with eczema over most of his body. He wasn't eating well and was losing weight at a frightening rate. Irene described how she felt when she realized that she hadn't had to buy Sean new clothes in over a year. "You know," she said, "On top of the eczema over nearly his whole body, I realized we were not needing to buy clothes and I thought, that's not good. It's not good when your child is not growing. He was three but he was wearing clothes for an eighteen-month-old . . . he was three years old and he hadn't hit thirty pounds. You could put your fingers around his upper arm," demonstrating by touching her thumb to her forefinger in a small circle.

Irene had a hard time telling the story without getting emotional. It was terrifying, she said, to have a sick kid and feel so helpless about it. "I went to the pediatrician and I said, 'Oh my gosh, what do I do?'" To her surprise, her regular pediatrician said, "'Oh, that's not that bad eczema. We can put some steroids on it.' I mean, he was covered from head to toe in a rash and he says, 'That's not bad'!" She went on, "We got a big Costco-size tube of steroids. That's what they prescribed. But, I went home and I started thinking about how your skin is your largest organ, and he's only three years old, and I know my husband has eczema to this day and he uses steroids, and I never really thought about it. But when you think about putting it on your three-year-old and them [the doctor] telling you, 'That's not bad eczema. He's fine. He's just fine.' And I said to them, 'But something else is not right here. He's small.' And they'd say, 'Well you're small and your husband's not that tall. You're 5'4" and he's 5'9" . . . he's a small kid. He's fine.' And I said, 'No . . . I had another child before this, and there's something wrong. He's off.'"

61

Irene talked about feeling helpless and angry that her doctor was not taking her concerns seriously. He never mentioned food, she said. In contrast, she described her first visit with Michelle as an experience that was "like finding someone who spoke my language. She listened to me. She said, 'OK, you think he's not growing. Tell me what exactly you think is wrong with his growth, his weight, and so on. What do YOU see?' This was not someone who said, 'He's fine.' I said, 'I'm finally in the right place!' And there began our food journey."

Michelle told Irene to get Sean tested for food sensitivities. His results showed he was sensitive to all kinds of things: eggs, almonds, gluten, dairy. Looking back on it, Irene realized that she was feeding Sean so many of the wrong foods and it was hurting him, "and he didn't know it and neither did I, because I was feeding him dairy and yogurt and eggs and then someone said, 'Let's get him off the dairy.' But I switched him to almond milk, but he's allergic to almonds. So it all got worse."

When Irene talked about Sean's food sensitivities, she called them allergies. But, she reminded me, his regular pediatrician wouldn't call them that. "They did antibody tests [on Sean] and they showed no allergies. But that was a test for serious allergies, she said—you know, the kind that make kids go into shock. But his problems were *like* allergies. His body wasn't able to digest these foods." Irene believed, like Michelle, that these sensitivities—immune provocations—were a piece of the puzzle that created cascades of dysfunction in Sean. When Irene finally eliminated the four things from Sean's diet that he was sensitive to and switched to only organic food, he started to eat again, and even to have a strong appetite. She was tearful when she described it. "You know, his symptoms started disappearing until they were largely gone. It is 95 percent better than it was. The most important thing is that he's eating! Improvement is not the word here . . . This was a child who would not eat and now he is a child who wants to eat everything."

Irene wasn't sure the improvements in Sean's eczema could be entirely attributed to eliminating GE foods. "It was likely the whole picture of foods, and maybe also the pesticides that he ate with them, that probably worked," she said. "When his gut started to work correctly because we eliminated the immune reactions, his eczema pretty much cleared up." The added bonus was that, when Irene herself began following the same diet as her son, her asthma improved dramatically. "You know, before, I had to use inhalers regularly. I ended up at the hospital several times a year." She described keeping a note by her bed for her husband in case she needed it in the middle of the night. It read, "Don't panic. I am having an asthma attack and need to go to the hospital now." "Sometimes, even with inhalers, it was not working. At the hospital, I'd get on the nebulizer. But you know, I don't even need to use my inhaler anymore, since I got off dairy. I went from needing it multiple times a day ... when I went to bed, when I got up, on rainy foggy days, you know ... when I was with the kids at practices, being outside all the time ... and now I only need it once a week or so. I mean they [inhalers] were everywhere. I wouldn't go anywhere without them."

Irene said she felt like her world had been turned upside down. How was it that none of the doctors she'd seen her whole life, and now none of her son's pediatricians, had ever mentioned that she or her son could be having bad reactions to foods or that the foods she was eating were tainted with pesticides? The most obvious culprit in her recovery was dairy. She eliminated dairy and her asthma went away. Over time, and many conversations with Michelle, she began to put it all together. "We're getting sick from the poor quality of food. We all shouldn't be drinking dairy. I'm a dairy drinker from way back. My dad drank one glass of milk at every meal. I love milk. But these people with dairy allergies, I thought, what do you do? But then I watched with my son. We had amazing results, just taking him off dairy ... And me and my husband, too. I would never have said something like [that we all should be off dairy] but here we are ..." I asked Irene to elaborate. She said the dairy we drink now is full of contaminants. It was also a GE food because it came from cows who were fed hormones made by genetic engineering. But this dairy also had pesticides and antibiotics. "Organics," she said, referring to how she avoids toxic ingredients in foods. "Only organics." There are "pesticides in the foods, genetic modifications of the foods, the antibiotics in the meats and dairy, and then all the other ingredients in the packaged foods. I read labels so carefully now," she said. "There is GMO corn, soy, wheat, even dairy in everything." (Wheat is not GE but, as mentioned, may contain lots of glyphosate.) "You have to read the fine print. I try to avoid those foods anyway, and to eat only organics. You can tell it's organic because the barcode number always starts with a nine."

Clients like Irene had been convinced by Michelle's approach and her theory that gut health required the elimination of glyphosate that they were getting in GE foods. The evidence was not just in the improvement in their children's health but, as with Irene, improvements in other family members' health as well. Irene didn't care what the experts said about the safety of GE foods or about the absence of standard guidelines in clinical practice for evaluating chemical load. She had all the proof she needed that, with help from Michelle, she was on the right path. She offered a series of observations

63

and intuitions about her child: his delayed growth and lassitude, his "being off," her shock at learning of his food sensitivities, her belief that Michelle was correct in her assessment of Sean's systemic food troubles showing up on the surface of his skin, in his behaviors, and in his physical decline. The pathology underlying Sean's health problems had to be sensed beyond these surface signs—in the interplay of physician and patient sleuthing, cobbling together a coherent theory with multiple lines of cause and effect, recognizing success as a personal, visceral experience of wellness, of having gotten to the root problems. Parents were often looped into their children's halo of disorder like this, with improvements in their own health as they altered their diet to help their children.

The patients who showed up for the first time at Michelle's clinic were usually very sick. One nine-year-old boy had ulcerative colitis, the culmination of a series of health problems that had begun soon after birth with an infected circumcision, several courses of antibiotics, an emergency trip to the ER with encephalitis, and surgery to put in a brain stent, followed by early and ongoing series of digestive troubles. He was on the verge of having his colon resected and was possibly heading toward J-pouch surgery, in which surgeons would remove large sections of his inflamed intestinal tract and redesign a new one. His mother told me his gut microbiome "never really had a chance." Michelle spent a year advising his mother about how to build up and repair her son's gut microbial environment by eliminating chemical-rich foods and loading up on probiotic foods and remedies. When I met him, he was nearly a teen, back at school and symptom free. His mom was not sure which of all the therapeutics actually led to the improvement—the homeopathic remedies, psychotherapeutic support, diet alterations, elimination of pesticides from his food, loading up on probiotics—but she said it was probably some combination of all of them. She was firm in her assessment that she didn't really think that all of Michelle's advice was worth heeding. Nor did she hold anything against the physicians who were involved in her son's early health crises—multiple courses of antibiotics as an infant, encephalitis and need for brain surgery, the advice that he "just eat more white bread." "They [conventional doctors] saved his life, twice," she said. Of course, things might have been very different for her son had he not started out with an infected circumcision. Here too, a successful clinical experience arises from sorting out and sensing the right path amid many options and choices.

Zoe was a six-year-old who lived with her parents in California's wine-growing region. She had first seen Michelle three years prior. Her mother, Debbie, described being overwhelmingly frustrated by Zoe's uncontrollable hyperactivity and rapid mood swings. Her mother described living with her as like living with Dr. Jekyll and Mr. Hyde. When Debbie was told by a pediatrician that Zoe should begin taking Risperdal, a medication used for patients with psychosis, she asked around and found out about Michelle.

Michelle's intake for Zoe lasted an hour and a half, during which she learned that in addition to her behavioral issues, she clearly had a lot of digestive or gut problems. She regularly had severe constipation and abdominal bloating. Michelle also learned that Zoe's family lived near an agricultural area, on the edge of many vineyards, and they had neighbors who were struggling with behavioral problems with their kids. She suspected environmental toxicants might be at work based on all these things: the behavioral and neuropsychological issues, constipation, bloating, and cycling moods. She encouraged Debbie to ask her neighbors to test the soils near where they lived; they found it full of toxic chemicals. She ran a toxicology screen on Zoe's urine and found elevated levels of petrochemicals, styrenes, plastics, and parabens. She suspected that if she could have tested for glyphosate, she would have found that too.

Michelle often pursued therapies that were intertwined with diagnostics. If you clear toxicants and the patient gets better, you have a better sense that the toxicants were a root cause. She recommended Debbie detoxify Zoe's environment. This meant getting rid of chemical-rich foods. It also meant getting chemicals out of the house by eliminating certain cleaning products and plastics. She encouraged Debbie to partner with other parents to convince the school and neighborhood to clean up the school and the leach field, both of which were unsuccessful. Michelle also worked on improving Zoe's ability to detoxify from unwanted chemicals using an amino acid protocol along with antioxidants and certain B vitamins that would improve cellular processes needed for detoxification and help the liver clear toxic substances through one of its key capacities, methylation. One of the things Michelle discovered from a genetic screen was that Zoe was homozygous for a mutated gene she needed to properly methylate—she was missing a properly functioning methylation system. This meant that whereas other kids in the neighborhood might be absorbing chemicals and eliminating them, Zoe was unable to. She couldn't repair the mutation, but Michelle reported that the elimination of toxicants and the addition of supplements that helped Zoe to detoxify resulted in a terrific outcome. When I met Zoe three years later, she was, according to her mother, really healthy—a world apart from the Zoe of three years prior.

I also met Mike, a sixteen-year-old who, when Michelle first met him at fourteen, was borderline obese and on several psychoactive pharmaceuticals

65

for episodes of extreme aggressive behavior (including episodes in which his parents called the police because he was threatening his mother with a knife and ones in which he had been forcibly removed from school). Michelle first worked to wean him off of the drug therapies he was on, collaborating with a psychiatrist to winnow them down to a few key treatments. Then she discovered his food sensitivities and removed those foods from his diet (especially peanuts but also cabinets full of packaged chips and gluten-rich foods). She convinced the family to switch his diet to only organic foods, and again worked closely with his psychiatrist to get him off psychoactive drugs altogether.

Within a year, Mike had settled emotionally and lost thirty pounds. But he often relapsed by going back to his "trigger foods" (junk foods loaded with sugar and chemicals), because the family reasoned that it was not fair for them all to be deprived of these foods if Mike was the only one with problems. They kept the pantry that had all those foods locked, but occasionally a brother would leave it unlocked and Mike would sneak in, grab, and consume a box of cookies or a whole jar of peanut butter privately in his room.

Michelle went around and around with the family, trying to convince them that they would all be better off eliminating these foods, but she got little traction. When I met the family, they admitted that they knew the changes in diet had made a huge difference in Mike's health—"It was a matter of life and death," Mike said—but they were not ready for the entire family to follow Mike's lead. Michelle eventually told the family she probably couldn't do much more for Mike if he couldn't stay off the trigger foods. Still, Mike's mom stressed how important it was to use only organic foods, "foods without pesticides," even if much of it was processed and packaged for long shelf life. Mike talked about how these foods just "made him go crazy." His father, who confessed to enjoying a jar of peanut butter with pancakes with Mike, realized that he, too, "did better off the packaged and processed foods." He just "felt better," he said, but he also lost weight and thought more clearly.

A three-year-old with chronic eczema, an eight-year-old with chronic migraines, a three-year-old with neurocognitive agitation, a sixteen-year-old with a collection of symptoms (chronic fatigue, blurred vision, partial paralysis, foggy brain, periodic mania)... these were the kinds of disorders Michelle strung together into a coherent and comprehensive theory of pathology—a coherent theory of health starting in the gut—that placed the most blame on chemical-rich foods. Even though her diagnostic sleuthing led her to consider other factors (a child with a faulty methylation gene, a

66

teen who also had tested positive for Lyme disease, etc.), the entry point for most of Michelle's diagnostics was GE foods, meaning glyphosate-rich foods. "Even if you find another culprit, like Lyme disease," she said, "getting rid of glyphosate will heal the gut and help get the immune system back to normal, [help get] digesting and the ability for the body to obtain nutrients back on track." This will help, she said, with even the most intractable diseases, even those, like Lyme, for which you might need conventional medicines.

Not all of Michelle's therapeutic recommendations worked. She often threw the "whole medicine kit" at them. This seldom meant using conventional pharmaceuticals, although it sometimes did. The constant for Michelle was the idea that food could be the root cause and cure of all of these disorders. Her convictions were not based on certainties offered in standard biomedical clinical guidelines. Nor was I able to say with much certainty that the improvements Michelle did see in her patients were primarily attributable to the removal of GE foods and glyphosate, as opposed to being an effect of changes in the diet more generally (such as eliminating dairy, gluten, or sugary junk foods), other remedies she used (including pharmaceutical and homeopathic ones), or some combination of all of these things. Some of her patients' improvements could have been based on what biomedicine would call psychosomatic effects, especially since Michelle was so good at listening to her patients and attending to their personal habits. But this did not stop Michelle from forming a very convincing global theory about the root causes of most of these health problems that kept GE foods and glyphosate at the center.

Her diagnostic and therapeutic choices were based on a complex weaving of the evidence at hand. She often told me, "I have to read the science myself." And she did. Lots of it. And, despite having no more training in these research fields than a typical physician would, she gleaned from these materials insights that helped her stitch together a coherent model for pathology. Making connections between foods, physiology, system and organ disruption, and chronic morbidity was not included in her medical training, but it worked. She focused on lines of causality that were circuitous. In the first instance, inflammation and leakiness of the gut were caused by eating GE foods, specifically by the activity of glyphosate, which is a known antimicrobial (meaning it could be killing off healthy gut bacteria). She believed the Cry protein from Bt-modified plants in GE foods was also a cause of gut inflammation. This inflammation from GE foods was what Arpad Pusztai found in his rat studies, she said; she had also found another researcher, Dr. Judy Carman, who had affirmed this reaction in pig studies.

67

An inflamed gut was a leaky or overly permeable intestinal lining, which meant molecules (including toxic chemicals) could get into the bloodstream that didn't belong there, chronically triggering the immune system and even passing these chemicals into the brain, where they inflamed tissues and disrupted synapses, causing mental health problems. Like an engine that was not turning over despite a driver's attempts to turn it on and flood it with gas, an immune system that was chronically provoked and then deprived of its cellular nutrition (notably minerals and amino acids) would cough and sputter and react but never mount a healthy response. An overactive yet underperforming immune system, she reasoned, might even result in autoimmune reactions, but at a minimum explained many of the skin rashes, foggy brains, and chronic diarrhea.

Michelle also argued that it stood to reason that because glyphosate was a chelator, it was binding important minerals such as magnesium, manganese, zinc, and calcium, depriving bodies of these essential ingredients for health. She also followed the research of geneticist Michael Antoniou at King's College London, who showed that glyphosate was harmful to liver and kidney tissues in study animals. Michelle argued that if glyphosate was having antimicrobial effects in plants, it might do the same in guts, where it was likely creating dysbiosis. Because glyphosate blocks the enzymatic pathway that microbes use to produce amino acids, this would also deplete the body of needed proteins. Pulling these threads together by reading the latest literature on leaky gut, dysbiosis, and the microbiome (Fasano 2012; Sonnenberg and Backhed 2016), Michelle was convinced that clinical science just had to catch up to what she was already seeing.

In fact, none of the microbiome research actually showed the specific routes to pathology that Michelle envisioned. These researchers were only beginning to understand the relationships between gut microbes, food, and pathology (Finlay and Arrieta 2016; Sonnenberg and Backhed 2016; NASEM 2018; Carmody et al. 2015; Lichtman et al. 2016; Spanogiannopoulos et al. 2016). Even though many researchers were pointing to strong links between gut health, brain health, and some foods (like high-fiber diets), and a few were talking about "leaky gut," there was little research on how chemicals beyond antibiotics impacted gut health (NASEM 2018), and virtually no research on the differences between organic and GE foods in relation to health or on the specific impact of glyphosate in the gut.

Michelle was not deterred by the absence of a scientific archive that pointed directly to the connections she made. She felt that the problem with waiting for clinical science to catch up with the facts coming out of more basic laboratory and toxicology studies was that industry funding would get in the way of, and in fact derail, these facts. Despite the growing body of literature that showed probable human health harms from glyphosate, pharmaceutical and chemical industry hacks would try to discredit this work. It was not hard to believe her about this. Still, she hoped that the more research that came out on the health effects of glyphosate, the more evidence would mount about its pathogenicity, and the harder it would be for industry researchers to conceal or whitewash these things. We'll see about that as we go forward in the next chapters.

Michelle's clinical theories were what some of her colleagues called "speculative," even if they translated into clinical practices that were often effective. But she maintained that she was working with the empirical evidence at hand that pointed to the obvious. Her go-to evidence of the harm from glyphosate-rich foods was one patient, a young boy who had kidney disease and whose father's kidney function was even worse than his. They lived in the Central Valley of California, surrounded by farms that were heavy glyphosate sprayers. She had been told by the father that his own kidney function was down to 20 percent when he brought in his son for a clinical visit. After the family switched to organic food to care for the son, both son and father got better, apparently restoring the father's kidney function to 80 percent. As far as Michelle knew, switching to organic food was the only remedy he pursued.

I like to think of Michelle as operating in and through her own sensorial swirl—using her skills and sensibilities to move from a weak evidence base to a commitment that pathologies could be made visible by removing triggers of harm and watching the results. She harnessed her data points and amassed a field of empirical evidence that sometimes drew from a selective scientific archive and other times from a patient reporting that they were getting better. This meant moving radically away from the standard guidelines for clinical practice that she had learned as an MD toward working through a swirl of evidence and carving certainty with a specific type of organization of the facts in relation to glyphosate. Glyphosate, and the foods grown with it, became key interlocutors for her clinical sensibilities; they enabled her to craft a theory of cause and effect that married the scientific evidence of chemical harm to leaky guts and the failing physiological and anatomical systems that cascaded into a multitude of symptoms and diseases in her patients. Glyphosate enabled Michelle to craft a clinical practice that made heterogenous clinical presentations seem like they were related, different outward manifestations of the same core etiological source.

69

The Clinical Swirl

What can patients like Sean, Amelia, Zoe, and Mike teach us about harm detection or attuning to the subtle changes in sensorium and capacity that have no medical diagnosis but are felt nevertheless? Habituation to chemicals asks us to think about how we inhabit chemicals, sometimes (but not always) in ways that enable those chemicals to come to rest in tissues and organs and cause disease. But not every patient who eats glyphosate-rich food will show the same signs of disease or dysfunction. Michelle often said, and Zoe's case suggested, that how much chemical load a patient can tolerate is a function of how well their methylation genes enable them to get rid of harmful toxicants. Sometimes even a body with a healthy methylation capacity will be overwhelmed by the sheer amount of chemical entering its tissues.

The seemingly arbitrary presentations of uneven effects of chemical exposures suggest uncertain yet compelling lines of cause and effect. Varied responses and a variety of different kinds of cascading ailments were to be expected, according to Michelle. Each patient had their own story of possible exposure and exacerbations. Sensitivities added to toxic food added to genetic predispositions added to exposure to parasites or other external sources of harm (like Lyme-infected ticks). Still, Michelle was able to carve certainty out of these bits and pieces of possibility by putting them in conversation with the science she read, the theories of harm caused by toxic foods, and her commitment to trusting what she saw. What made her clinical practice effective was her ability to sense the evidence of chemical harm and cultivate a sensible route through the evidence toward a theory of pathogenesis. Chemicals enter and swirl in bodies as consequences of long-term infrastructures, food systems, and regulatory processes. Relying on a scientific swirl of evidence offered Michelle moments of certainty and conviction that could bend and flex with the nonuniform presentations of harm, also enabling these chemicals to serve as a stabilizing force, for a theory of causal relationships and an innovative form of clinical care.

Glyphosate's interabsorptions and mixings give rise to theory-making and the building of clinical attunements. If we follow Michelle's thinking, glyphosate moves through bodies and settles in all kinds of tissues, forming clusters of impact in relation to processes that are needed for health. Once in tissues, glyphosate can effect alterations that could be tolerable under some circumstances, or not. It could even settle in ways and in some tissues (like genetic tissues) in ways that could be deadly. The body is, in

Michelle's view, a multiplicity of physiological systems that are in constant motion, and these systems (and the material tissues that constitute them) can reveal the presence of the chemicals. Like Deleuze and Guattari's body without organs (Deleuze and Guattari 1987), I think that glyphosate attunes us to the ways in which chemicals swirl in bodies, alighting in an organ here, a collection of tissues or molecules or cells there, territorializing around one kind of disruption in one person, a different kind in another, augmenting the presence of glyphosate in ecosystems of potential bodily transformation. Michelle's clinical practice formed a sensorium of operations that exposed the effects of chemicals, that apprehended the varied presences of glyphosate in ways that mimic the momentum and movement of the chemicals themselves, forming a swirl.

The facts and the clinical perceptions that Michelle brings to bear on behalf of her patients reveal an attunement to and a comfort with the perpetual motion of different formations of the swirl. Rather than saying this is a consequence of an integrative medical approach, I would suggest that efforts to trace chemical harm must, in some sense, follow the chemicals themselves, tracing their potencies and differential impacts on different tissues and bodies and producing different kinds of symptoms across populations. Chemicals, in this way, live through people, moving in uneven ways through the body, settling in some tissues and not others, disrupting absorption here and clustering as mutations there. Glyphosate is like this; it swirls and settles and enables the formation of different kinds of certainties that, in turn, compel one kind of therapeutic action over another.

The swirl of glyphosate is but one of the numerous chemicals we could talk about in relation to food-related toxic exposures. Many agrochemicals are widely used in the United States, revealing agricultural decisions that most people have had little to no control over, no way to opt out of. But glyphosate offers a particularly useful example of how chemical life can seep into other domains, far beyond agriculture, including clinical spaces where physicians like Michelle attempt to capture and harness its multiple effects. Much like a body that has been absorbing toxic chemicals for many years, mutating tissues and cells in fluid and surreptitious ways that may or may not be causing a rash here, diarrhea there, headaches here, kidney failure there, glyphosate offers a stabilizing constellation of possible explanations for bodily harm and moments of enough certainty to effect therapeutic interventions. Glyphosate, in this sense, is a good interlocutor for how we might not just live with chemicals but also think about them in relation to knowledge and action.

71

In many fields of action—from clinics, regulatory industry testing companies, and EPA panels all the way to websites where the safety of food pesticides are contested—I have had to become attuned to, and sense-aware of, how chemicals like glyphosate keep things unstable. This is particularly visible when we turn back to the scientific archives and activist efforts around this chemical. To start with, then, I turn in the next chapter to the disruption glyphosate has caused in deciphering a scientific consensus.

Chapter Four

So, it comes to this: we must trust our scientific experts on matters of science, because there isn't a workable alternative. And because scientists are not (in most cases) licensed, we need to pay attention to who the experts actually are—by asking questions about their credentials, their past and current research, the venues in which they are subjecting their claims to scrutiny, and the sources of financial support they are receiving. If the scientific community has been asked to judge a matter—(as the National Academy of Sciences routinely is)—or if they have self-organized to do so (as in the Ozone Trends Panel or the IPCC), then it makes sense to take the results of their investigations very seriously.... Sensible decision making involves acting on the information we have, even while accepting that it may well be imperfect and our decisions may need to be revisited and revised in light of new information. For even if modern science does not give us certainty, it does have a robust track record.... While these practical accomplishments do not prove that our scientific knowledge is true, they do suggest that modern science gives us a pretty decent basis for action.—Naomi Oreskes and Erik M. Conway, *Merchants of Doubt* (2011)

The Scientific Consensus & the Counterfactual

When glyphosate prompted Monsanto scientists to create a new kind of food by modifying plant genes such that they could live with its deadliness, it also brought into existence a world of trouble in relation to evaluating the possible reach of this deadliness to humans. In chapter 3 I described this trouble as partly a consequence of glyphosate's ontological multiplicity. In chapter 4 I explored how glyphosate's multiple potencies prompted new clinical sensoria, enabling physicians and patients to adjudicate health in and through plausible links between evidence of injury and known potencies of this chemical. In this chapter, I return to the question of how glyphosate has become a key interlocutor in the arbitration over food toxicity. The facts about its potencies, often contradictory, are multiplied in the journals and archives of science written by those who care about it. In the scientific archive, glyphosate has provoked a veritable war over the truth, wreaking havoc with something called the scientific consensus.

Like many a wayward toxic chemical, glyphosate has produced in the scientific archive conflicting claims about harm (Fortun 2012a, 2012b; Boudia 2014; Davis 2014). The contested facts about chemical harm in general arise from the way chemicals themselves move in and through environments and bodies in diverse ways that are hard to trace (Murphy 2006, 2008; Jain 2006; Landecker 2011; Nading 2016). Moreover, the science can be unreliable in relation to toxic chemicals in the environment for other reasons. As Boudia and Jas (2014a) note, scientific accounts of carcinogenicity using rubrics of allowable dose levels developed in the 1950s and 1960s were supplanted by notions of "socially acceptable risk" that took into account political and economic considerations regarding the regulation of their use. This meant that the goal of regulation moved from assessment to the development of reasonable policies about managing toxicity without eliminating the chemicals themselves. In other words, the move was toward incorporating into the regulatory process questions not just of toxicity but also of social benefit even if it came with certain costs, like increased risk to consumers or those exposed and the cost to industry profits. Finally, technologies used by regulatory agencies to evaluate toxicity tend to use reductionistic models that cannot capture the full scope of any chemical's impacts in the world (Murphy 2006; Boudia 2014; Krimsky 2014).

Scientific studies of environmental toxicity by regulatory agencies have not been immune to the rise of activism and other forms of counterknowledge production (Boudia and Jas 2014a), as well as by the rise of advocacy platforms that have crafted different narratives about how to regulate and identify toxicity (Roberts 2014). This has been especially true for the case of toxic pesticides (Davis 2014; Guthman 2019). These accounts explain a great deal about "the limits of scientific knowledge in resolving issues raised by toxicants and the often political nature of decisions regarding these substances" (Boudia and Jas 2014a, loc. 490), which lead to the production of ignorance, uncertainty, and profound asymmetries between various constituencies invested in eliminating or using them (see especially Frickel and Edwards 2014; Proctor and Schiebinger 2008). We see all of these processes, and more, at work in the story of glyphosate.

From its early celebrity as a discovery supposedly as important for reliable global food production as penicillin was for battling disease, glyphosate has now become a new kind of superstar. It is one of the most contested chemicals of the early twenty-first century, holding an outsize power in fields far beyond the industrial farmlands where it is sprayed regularly and often. One of the reasons it holds this status is its relationship to what is often called the *scientific consensus*. To trace the ways that science is being shaken up by glyphosate—giving rise to a swirl—requires also delving into the various ways it has been involved in forming scientific consensus.

The Scientific Consensus

The scientific consensus is a strange beast. It is not a fixed thing with eternal, unalterable qualities handed down at one point in time by regulatory agencies or scientists (Oreskes and Conway 2011). Any scientific consensus is always being created and recreated by scientists in different locations at different times in response to questions about the certainty of scientific facts. The scientific consensus is, in other words, an artifact of the need for such a consensus in the face of persistent dissent, controversy, or uncertainty about the facts. And yet the idea of a consensus is that it can settle the facts in specific ways. A consensus can be used as a touchstone for policy makers, activists, clinicians, and any constituency operating in spaces where certainty is needed. That is, by saying that the consensus is produced in and by contestation, I do not also mean to say that the consensus does not traffic in dreams of a fixed certainty. These dreams of certainty are brought to life by the scientific communities that construct them. Thus, despite the fact that most scientists would say that their research can provide only provisional truths that are limited to the conditions and contexts of their investigation, something called the consensus is still thought to be dredged from these provisional practices and that consensus is considered imperative for decision-making. To trace this in relation to glyphosate, I return to the case of GE foods and how they have been deliberated in and through a variety of scientific infrastructures. To be sure, the scientific consensuses about the safety of GE foods and glyphosate are not exactly the same, but both have been produced by the same assemblages of scientific activities. The two have, in many ways, become inseparable.

Although challenges to scientific certainty about the safety of GE foods began almost immediately after they were invented, being what Schurman

75

and Munro (2010, 175) call a "permanently contentious technology" (see also Weasel 2009; Stone 2015), there is currently a scientific consensus that says both GE foods and glyphosate are safe (NASEM 2016). This consensus is repeatedly used (and named as such) to discredit widespread public concern that these foods and pesticide companions are not safe. What is this consensus based on?

The most recent scientific consensus on GE food and pesticide safety is presented in a metastudy report titled *Genetically Engineered Crops: Experiences and Prospects* funded by the National Academies of Science, Engineering and Medicine based on a collaboration that began in 2014 (NASEM 2016). Although the report is now somewhat old in the timeline of scientific publishing, it was not the first time that a metastudy was used to suggest a consensus on the safety of these foods and their associated pesticides nor would it be the last (Williams et al. 2000; Andersson et al. 2005; Nicolia et al. 2013), and it remains the most commonly cited document among lay publics and scientific communities who make claims about the scientific facts about these foods and chemicals today. This is true despite the presence of a lively and robust counter-archive on the risks of these foods and glyphosate, including the arrival of a controversial and damning report in 2015 about the specific carcinogenic health risks of glyphosate, which I will get to. Here I spend some time on the construction of the NASEM report itself along with its findings.

Readers of the NASEM report are told at the outset that the thirty scientist-authors of the report chosen by the National Academy all have blue-ribbon credentials, meaning not only that the scientists are established experts but that these experts are not guilty of the single most often used tactic for discrediting the science: that they were paid by industry. The report is 606 pages that weave a tapestry of information about comparative results of hundreds of studies to decipher where the preponderance of evidence points. The authors rely on research that uses different methodologies, different kinds of expertise, different objectives, and different objects of study. The report is not short on thoroughness, and it deserves a substantial amount of attention not just for its truth claims but also for how it makes them. Glyphosate travels through the pages periodically as a singular concern. But, for those specifically interested in glyphosate in relation to human health (as opposed to its impacts on environments, livestock, or poultry), it helps to go straight to the eighty-page chapter titled "Human Health Effects of Genetically Engineered Crops." That section, based on the evidence presented in three hundred studies available by 2014, starts by offering a note about how

76

to read the report that rehearses many of the same concerns I have listed already in this book:

> The committee thinks that it is important to make clear that there are limits to what can be known about the health effects of any food, whether non-GE or GE. If the question asked is "Is it likely that eating this food today will make me sick tomorrow?" researchers have methods of getting quantitative answers. However, if the question is "Is it likely that eating this food for many years will make me live one or a few years less than if I never eat it?" the answer will be much less definitive. Researchers can provide probabilistic predictions that are based on the available information about the chemical composition of the food, epidemiological data, genetic variability across populations, and studies conducted with animals, but absolute answers are rarely available. Furthermore, most current toxicity studies are based on testing individual chemicals rather than chemical mixtures or whole foods because testing of the diverse mixtures of chemicals experienced by humans is so challenging. (NASEM 2016, 171–72)

Drawing readers into both the commitments and disclaimers about certainty can help produce confidence in the consensus, but it also might not. Critics of GE foods are likely to note that this disclaimer, like that seen for any chemical harm, is also a device for obscuring possible harms from these foods and pesticides. As I mentioned in chapter 1, evidence of chemical harm is easily diffracted onto a multitude of possible sources, exculpating many a guilty chemical. The dubious reader might also note that although no scientist was paid by industry to write the NASEM report, it says nothing about whether these scientists' own research was ever funded by industry sources, nor about whether the studies they rely on to produce the metastudy were mostly funded by industry or industry-funded academic institutions.

Nevertheless, to the sympathetic reader, the admission of limitations might work to affirm the trustworthiness of the report by reminding readers that concerns about industry meddling directly or indirectly in the findings should be laid to rest and that the authors will not be overstating the findings, just sticking to the known facts despite their limitations. In this report, a lack of absolutes appears as a reminder that science in general is rarely able to produce absolutes about anything—which is not exactly the same thing as saying it thrives on uncertainty. What it can offer, however, is an abundance of evidence, and that evidence can point to enough reasonable conclusions and convictions to suggest there is a consensus.

77

Consensus is forged by other means as well. Not many readers of the report will have enough scientific expertise to make astute evaluative judgments about the quality of the work in each study, let alone what their conclusions actually are. This is also true of many of the scientists who construct it, who usually have expertise in one but not all areas under consideration (not unlike the limited expertise of physicians). Thus one can, as I believe many do, just read the summary or the brief take-home points offered at the end of each chapter. This shorthand route to the findings is a time-honored tradition in the scientific world, where it is assumed that peer review by experts will deter poor-quality scientific research from reaching the light of day and thus that the summary results can be trusted. The NASEM results are produced by experts who have presumably read not just the summaries of each study but the entirety of the studies, scrutinizing their methodologies and findings carefully. Nevertheless, readers of the NASEM report will come away from the report with an entirely different sense of the facts than if one were to read it in detail or, better yet, return to the original studies it relies on. I won't do that here, though I will go into some specific details about what the report claims (the report is easily accessed on the NASEM website).

The summary of the chapter on health effects offers the consensus view by telling us that there are three ways to test GE crops and foods derived from them: animal testing, compositional analysis, and allergenicity testing and prediction. It reminds us that there is little evidence that animals fed GE foods are harmed by them. It states that although one can find statistically significant differences in nutrient and chemical composition between GE and non-GE plants, this variation is no greater than that which can be seen in normal variation in all non-GE plants. It then offers the specific assurances about health:

> The committee received a number of comments from people concerned that GE food consumption may lead to higher incidence of specific health problems including cancer, obesity, gastrointestinal tract illnesses, kidney disease, and such disorders as autism spectrum and allergies. There have been similar hypotheses about long-term relationships between those health problems and changes in many aspects of the environment and diets, but it has been difficult to generate unequivocal data to test these hypotheses. To address those hypotheses with specific regard to GE foods in the absence of long- term, case-controlled studies, the committee examined epidemiological time-series datasets from the United States and Canada, where GE food has been consumed since the mid-1990s, and similar datasets from the United Kingdom and western Europe, where GE food is

not widely consumed. The epidemiological data on some specific health problems are generally robust over time (for example, cancers) but are less reliable for others. The committee acknowledges that the available epidemiological data include a number of sources of bias. (NASEM 2016a, 9–10)

It then concludes that there is no evidence of "differences between the data from the United Kingdom and western Europe and the data from the United States and Canada in the long-term pattern of increase or decrease in specific health problems after the introduction of GE foods in the 1990s" (NASEM 2016a, 10). It specifically states there is no evidence of differences in rates of cancer, obesity, type II diabetes, chronic kidney disease, celiac disease, autism spectrum disorder, or allergies that can be tied to consumption of GE foods.

From the three hundred research articles that were put into conversation with one another as comparative points of reference come answers to specific questions about evidence of harm, or rather the lack thereof. And yet, in many ways, the report's ability to claim a consensus depends as much on how it manages the disparate kinds of information as it does on the actual data in the studies. It is worth noting that the specific portions of the NASEM report dealing with glyphosate are crucial to the conclusions reached in the study.

The first thing one reads in the substantive part of this chapter in the NASEM report is a boxed list of passages from studies that have shown that these foods and their pesticides are safe—another tactic of consensus-building by noting the many reputable organizations and scientific institutions that are in agreement:

Sample of Statements about the Safety of Genetically Engineered Crops and Food Derived from Genetically Engineered Crops

"To date, no adverse health effects attributed to genetic engineering have been documented in the human population." NATIONAL RESEARCH COUNCIL (2004)

"Indeed, the science is quite clear: crop improvement by the modern molecular techniques of biotechnology is safe." AMERICAN ASSOCIATION FOR THE ADVANCEMENT OF SCIENCE (2012)

"Bioengineered foods have been consumed for close to 20 years, and during that time, no overt consequences on human health have been reported and/or substantiated in the peer-reviewed literature." COUNCIL ON SCIENCE AND PUBLIC HEALTH (2012)

"[Genetically modified] foods currently available on the international market have passed safety assessments and are not likely to present risks for human health. In addition, no effects on human health have been shown as a result of the consumption of such foods by the general population in the countries where they have been approved." WORLD HEALTH ORGANIZATION (2014)

"Foods from genetically engineered plants intended to be grown in the United States that have been evaluated by the FDA through the consultation process have not gone on the market until the FDA's questions about the safety of such products have been resolved." U.S. FOOD AND DRUG ADMINISTRATION (2015)

"The main conclusion to be drawn from the efforts of more than 130 research projects, covering a period of more than 25 years of research, and involving more than 500 independent research groups, is that biotechnology, and in particular GMOs are not per se more risky than, e.g., conventional plant breeding technologies." EUROPEAN COMMISSION (2010), QUOTED IN NASEM (2016, 172)

Before summarizing all the studies used in the report, collated and grouped together for general conclusions about the preponderance of evidence, the report explains the reasoning that led these authors to their conclusion that GE foods and pesticides are safe. The first, reminiscent of arguments made in the deliberation over DDT, is a discussion of the potential for naturally occurring toxicants in food crops, including various endotoxins that can have protective effects against herbivores and pathogens that could harm plant survival. The idea here is that even naturally occurring foods can be toxic, so GE foods should be measured not against all non-GE foods, but rather against the potential toxicity of other non-GE foods: are they any more toxic than some of these? This leads to the conclusion about "substantial equivalence," noting, as I said in chapter 3, that GE crops pose no more or less risk than any other plant bred naturally or agriculturally for potential endotoxicity.

This is followed by an extended discussion of the idea that Bt crops might pose special risks, but the studies used in the report do not show this (though some of the studies could have used better methods). The NASEM authors also remind readers that US regulatory agencies take all these studies and others that explore the admixture of herbicides with Bt into account when establishing their safety as pesticides. The report then describes the typical methodologies for studying toxicity of these foods, and particularly glyphosate, in laboratory animal studies. In part as a gesture to evenhandedness and lending the report credibility, the report offers a highlighted text box on pages 188–89 about the work of Gilles-Éric Séralini and his lab, including the findings of his research that suggested toxicity from glyphosate, and the criticism it received. The report titles this section "Controversial Results of an Animal Feeding Study of Genetically Engineered Crops and Glyphosate."

I will return to Séralini's work below. Briefly, Séralini published results of studies from his lab that showed GE foods and Roundup (glyphosate) were probably carcinogenic. The NASEM report focuses specifically on Séralini's two-year feeding study of rats, the results of which were published, then subjected to forced retraction, then republished, only to face more criticism from the European Food and Safety Authority, which reread the raw data of his study results and argued that they were not indicative of a statistically significant difference (EFSA 2012). The NASEM authors note the EFSA (2012) critique of Séralini's interpretation of results and offer their own additional conclusion: "It is important to note that all the bars for the female control groups are the same because the same ten female rats are always being compared with the different treatment groups. If only three of the female control rats had one extra tumor, the graphs would show no differences" (NASEM 2016, 189). This is why, they remind us, that "reanalysis of the data (EFSA, 2012) found no statistically significant differences" (NASEM 2016, 189). In other words, they are suggesting that Séralini overstated the tumorigenesis findings based on their and others' rereading of his results. For now, I note that the labeling of the highlighted Séralini portion of the report as "controversial" seems meant to make it easier to agree with the committee's own conclusion: that we should all disagree with Séralini's conclusion that this one study should "lead to a general change in global procedures regarding the health effects and safety of GE crops" (NASEM 2016, 189).

The NASEM authors also point to four other studies that followed Séralini's research design but produced opposite results (Magana-Gomez and de la Barca 2009; Domingo and Bordonaba 2011; Snell et al. 2012; Ricroch 2013) and conclude that, based on these studies, there is no evidence of increased tumorigenesis from ingestion of GE foods or Roundup. Specifically, the report says, "Some found no statistically significant differences, but quite a few found statistically significant differences that the authors generally did not consider biologically relevant, typically without providing data on what was the normal range" (NASEM 2016, 194).

Often the deliberation over how risky GE foods and pesticides actually are comes down to a type of semantic gloss on the interpretation of research results and the use of the phrases "statistical significance" and "no biological significance." Anticipating concerns from critics who say that the idea of something having no biological significance is arbitrary may be a way for industry bias to rear its ugly head and obscure risk (Robinson et al. 2015), but because the NASEM authors are relying on many studies that offer this conclusion, they include this deep dive into this particular phrasing, focusing on whole food studies done by the European Food and Safety Agency:

> EFSA also published a document that focused specifically on the questions, What is statistical significance? and What is biological relevance? The accessibly written document makes clear that the two are very different and that it is important to decide how large a difference is biologically relevant before designing an experiment to test a null hypothesis of no difference. The problem in most whole-food animal studies is in determining how large a biological difference is relevant. Most of the statistically significant differences observed in the literature on the animal-testing data were around a 10- to 30-percent change, but the authors do not give detailed explanations of why they conclude that a statistically significant difference is not biologically relevant. A general statement is sometimes made that the difference is within the range for the species, but because the range of values for the species typically come from multiple laboratories, such a statement is not useful unless the laboratories, instrumentation, and health of the animals were known to be comparable. Clearly, the European Commission relied on both expert judgment and citizen concerns in making its assessment of biological relevance of the effects of GE foods in requiring 90-day testing. (NASEM 2016, 193)

In other words, despite the lack of sufficient discussion of the rationale for claiming something is not biologically significant other than the possibility that the statistical differences were based on different laboratory studies (that were perhaps not comparable), the NASEM authors are suggesting we should trust the experts doing the studies because they know what they are doing and have consumers' concerns in mind. The nuance of this kind of reasoning, in a report that itself is offering a metastudy that compares research from multiple different laboratories, should not be missed.

The rest of this section of the report offers a long list of other studies that show no significant biological differences in studies of livestock and poultry fed GE food, including Walsh et al. (2011), which is used to counter the work of Judy Carman (2013), who, as we will see, reports significant biological findings of intestinal inflammation in pigs. They use the work of a livestock biologist, Alison Van Eenennaam, who showed that measures of body and organ weight, growth measures, and veterinary dependency (all things that matter to commercial livestock industries) can serve as proxies for overall health effects (Van Eenennaam et al. 2007; Van Eenennaam et al. 2014). Most important, Van Eenennaam's work showed no higher rates of carcinogenesis in livestock fed GE food than control groups eating non-GE food and no pesticides.

Before I delve into how research studies are packaged for comparison to create a preponderance of evidence in one direction, including careful adjudication of each study's various strengths and weaknesses, I turn to the most important section of the report, meant to assure readers that GE foods and pesticides have no ill health effects on humans—that is, the epidemiological evidence. The primary comparison in this section is between the United States and Canada (where rates of consumption of GE foods are presumed to be high) and the United Kingdom and Western Europe (where rates of consumption of GE foods are presumed to be low because of regulatory efforts to slow their spread). I will not parse the details of the studies here (and again refer interested readers to the report), but the conclusions are compelling.

The authors first note that rates of cancer are on the rise in both study groups but they are not significantly higher in the United States than in Europe or the United Kingdom. Nor is there evidence of increased allergenicity to food (in humans) in either region. Nor, finally, is there epidemiological evidence of higher rates of kidney disease, obesity, gastrointestinal tract diseases including celiac disease, autism spectrum disorder, perturbations of gut microbiota, or transfer of transgenic DNA to animal somatic cells. A critical reader might argue that the fact that rates of all these disorders have increased in all of these places should be cause for alarm, but that aside, the point is that the rates are no higher in the United States and Canada than in the United Kingdom and Western European nations. Again, multiple epidemiological studies were used, including those used to create national archives on rates of prevalence of these various disorders.

The section ends with another gesture to evenhandedness: a description of the benefits to human health of GE crops and their pesticides, including benefits that could come from improved nutrient content of food through enhancement technologies, improved quality of fats (in rapeseed, also known as canola, for instance), reduced concentrations of toxins (specifically, a potato that can be fried without producing asparagine), and a reduction in farmers' exposure to toxic pesticides (referring to the use of Bt in plants). This is a very interesting section that reproduces almost exactly what one finds in industry publications and websites about the benefits of GMO technologies for food. Focusing on enhancements—most of which are not in commercial use or have not been shown in commercial use to be very

83

beneficial, such as vitamin A–enriched rice—serves to efface the reality that roughly 99 percent of GE foods are not designed with enhancements but are just Roundup Ready and Bt commodity crops.

The confidence of the health safety of GE foods and glyphosate in the NASEM report based on the epidemiological inventory of health conditions is one of the more interesting components of the report. The presumption that the United Kingdom and Western European countries are a useful comparator population to the United States and Canada is based on the fact that there was more public resistance to GE foods in the former countries and thus more regulation and less consumption of them. The careful reader who goes to sources beyond the NASEM report might be surprised to find that the only actual ban on GE food cultivation and sale in Europe was from 1999 to 2004 and did not include crops that arrived before 1999. As of 2012, forty-eight genetically engineered foods have been in circulation in the European Union in animal feed, imported food, food additives, and maize (Royal Society 2016). Furthermore, glyphosate was not banned in the European Union until after 2015 (Hessler 2020). It is true that in the United Kingdom, GE foods are subjected to stricter regulations about labeling and monitoring, but neither GE foods nor glyphosate has ever been banned in the United Kingdom; banning glyphosate is only now being debated there. I also remind readers that different nations of Western Europe follow different regulatory policies regarding the sale of GE foods, and in most cases the regulation is only about labeling, not preventing them from commercial sale (Hilbeck et al. 2020). As I mentioned earlier, a good story about the belligerence of farmers who are committed to using glyphosate comes from Italy, where a village that went glyphosate- and GE-free was repeatedly exposed to terrorist spraying of its crops with Roundup by (presumably) neighboring farms that used it (Ackerman-Leist 2017). Even where GE seeds are banned, farmers still use glyphosate-based herbicides. In other words, not much attention is given to the sleuthing of how much GE foods are actually being consumed in Europe and the United Kingdom (including by livestock), nor how much glyphosate is actually being used in these places.

One might want to quibble over the other limitations of the findings used in this report, including the fact that a new report from 2015 on the carcinogenic impacts of glyphosate (IARC 2015) was available but perhaps not soon enough to alter the NASEM report that was in press. Nevertheless, the NASEM report continues to serve as a foundation for most claims of there being a scientific consensus concerning the safety of glyphosate and GE foods. I want to spend a little more time, then, on the ways consensus has been forged in and through this scientific endeavor, since that effort has produced both a consensus and, as we will see in relation to glyphosate, a variety of other consensus claims.

Scientific consensus is made by the coalescing and collating of evidence in ways that allow metastudy authors to say that the preponderance of evidence points one way and not the other. One notable characteristic of this kind of work is *shingling*. That is, some of the studies cited in the NASEM study are themselves metastudies that use primary data from studies that appear again in the NASEM study (Nicolia et al. 2013; Williams et al. 2000). This layering of original studies, the metastudies that cite them, and the metastudies that cite the metastudies gives one the impression that a preponderance of studies all point to the same conclusions, even though some are simply repetitions of the same studies. Even the initial gray-boxed list of agencies and institutions that have affirmed GE food safety is based on a very small number of the same studies, but the layering of reference points to those who restate the views of these studies becomes a way of making the results seem much more robust than they would if it were clear that they all relied on the same bodies of evidence.

Another procedure that is used to create consensus is the scale-up of knowledge claims from studies that have little in common by putting them into conversation with one another in ways that move swiftly from the microdetails of their work and methods to the meta-insights from which metastudy conclusions are crafted. Data generated from a rat or mouse study here and a soil study there, an epidemiological study here and a livestock study there, enable study panelists to offer some general certainties about trends. Comparing apples to oranges, fitting square pegs in round holes, these free-floating data sets are put in conversation with one another in ways that accumulate momentum and point to trends that may or may not be consistent with the specific facts of the original studies.

The extraction processes of metastudies dislodge bits of evidence from study contexts that were quite different (agrosciences, environmental sciences, biomedical genetics, toxicology labs, epidemiology, and public health). Gathered up and set in conversation, the details of studies are erased: where and how they came into being and produced co-constituted forms of evidence, whether in laboratories, on farms, in chemistry labs, and in animal and human bodies; with industry funding or without it. The integration of the metadata is made to seem certain despite the conditions and contingencies of production or even the caveats found in the limitations section of each study. Thus, who paid for a study, what product lines it enabled, who

85

profited and continues to profit from it, the kinds of laboratory animals used and the limits of their experimental models, where the research took place and where their funding came from, and so on are all effaced, even though such details might alter one's understanding of the data itself. This is all before we ask questions about who sits on the editorial boards of the publishing houses where this work is reified as reliable enough for use in a metastudy in the first place.

Metastudies use scaling-up practices that erase the specific ways in which each study defines its limitations or leaves open questions of reading the evidence otherwise, making the facts feel tangible and worthy of consensus certainty. This scale-up, in other words, involves reducing the studies to their most reliable conclusions while simultaneously expanding them by setting them in conversation with other study results in order to form a claim about a preponderance of evidence. The way that reports like this are able to discern certainty from diffracted patterns of evidence requires forging certainty from an array of disparate facts, placing all of them together into a pattern that presents them as a crowded (indeed evidence-rich) basis for consensus—another way of making the research larger than its individual and specific claims. These processes rely on a compelling kind of citational practice that, as scholars have long noted, forecloses debate about facts and augments those that garner large social networks (Latour and Woolgar 1979).

One might note that this is how all scientific consensuses (if not all scientific works) are built (Oreskes and Conway 2011), so it is not surprising that this is how it has worked with GE foods and pesticides. There can never be one definitive study that can answer all of the questions one might raise about the ways glyphosate or GE foods might harm (or not harm) human health. Still, consensus is built not just by way of the amplification and collation of some facts and conditions so they appear more robust than they could individually but also by the silencing of certain facts and conditions that do not scale well. Séralini's work did not scale well; it had to be set aside and discussed outside of the scale-up. In this sense, a consensus offers a specific kind of certainty that can easily be made unstable not just by critical scrutiny of the details in the evidence base (which any number of critical readers might venture to do) but also by the very practices that are relied on to forge a unified, unidirectional set of claims. The formation of a consensus, in other words, relies on a process that diffracts the conditions of production of the research and the possibility that, when considered at that level of detail, would make things noncomparable. Consensuses must work hard to scale up—drawing together the facts in particular ways and then making them work in tandem with one another to forge a unified field of vision that suggests the possibility of certainty. And yet, as we will see, when it comes to the question of safety or harm from glyphosate, scale-up can also fail. This is more than saying, in the spirit of Thomas Kuhn, that evidence begins to unseat the prevailing scientific paradigm. What I am suggesting is a different kind of scientific architecture that has to persistently manage the counterevidence by practices of scale-up. To show how scale-up and paradigm thinking fail, I turn now to the counterfactual archive.

The Counterfactual Archive

The contestation over the safety of GE foods and their associated pesticides emerged simultaneously with their appearance in the world. This began when a few scientists who worked on the development of these foods and pesticides pushed for more scrutiny of them or offered overt warnings about them based on their research findings (Martineau 2001; Druker 2015). What is interesting about these scientists is not so much that they produced evidence of risk from these foods and pesticides but rather the way that they were so quickly and uniformly dismissed. Their work forms a robust archive that anyone sleuthing the history of glyphosate must wade through, reading carefully around the watermarks indicating publications that were forcibly retracted and separating hyperbolic accounts of industry-funded slander from claims about the quality and trustworthiness of the scientific work. These stories will be familiar to anyone working in the non-GMO activism world, but to the uninitiated, hold on for a wild ride.

For instance, Arpad Pusztai, the scientist who sparked Michelle's skepticism about GE food safety, is a Hungarian-born nutrition and biochemistry research scientist who features prominently in the history of the counterfactual archive on GE foods and pesticides. He was working at the University of Aberdeen's Rowett Institute in 1995 when he was asked by Scotland's Ministry of Agriculture to evaluate a new potato that had been genetically modified to be resistant to aphids. According to Pusztai, he did not have a predisposition about its potential harm. He reported that he began his research and that "we were all enthusiastic about it [the potato]. I was enthusiastic about it. The Ministry thought that if we did this study, looking at all aspects, then it would be an endorsement of GE. And, when they introduced it, it would say that the foremost laboratory in Europe—nutritional laboratory—looked at them and they found them alright" (Robin 2008, 00:50, 00:54 minutes).[1]

Pusztai's expertise was in potato lectins, naturally occurring proteins found in plants and animals. His research focused on a specific transgenic process using a lectin called the snowdrop lectin found in the snowdrop, a flowering plant (*Galanthus nivalis*, or GNA). The GE potatoes Pusztai studied were modified with the GNA lectin that they believed would enhance the potato's resistance to aphids. Before the research started, Pusztai verified that in their naturally occurring state, these lectins did not themselves pose a health risk to test rats. Then he tested the genetically modified potatoes that contained the lectin by feeding them to rats and comparing those rats to a control group that did not consume the modified potatoes. Pusztai described what he found:

> Twofold effects. First it [the GE potato] started to increase a proliferative response in the gut. And that you don't like because, um . . . this is possibly . . . I'm not saying that it is cancerous. But what it does . . . it can have an adjuvant effect on any chemically induced tumor. The other thing is that the immune system certainly got into high gear. We don't know whether it is good or bad, but it certainly did recognize the GE potatoes as alien. And we were convinced that it was the insertion [of the gene] that was the problem and not the transgene [the snowdrop lectin protein]. As I said, the transgene, when we did it in isolation, even at 800-fold concentrations, didn't do any harm. It was a very important point because the American FDA is going on about [this being] a *neutral* technology. And what we did say, and what we did publish and what we corroborated and confirmed, was that, it was not the transgene which was the problem, but it was the technology [modifying the genes in this way]. (Robin 2008, 1:50, 1:54)

Pusztai said the rats fed the GE potatoes had intestines that were nearly double the size of those of control rats, an indication that the rats' immune systems were reacting to the food, causing inflammation. The rats who were fed the lectin protein alone were not affected, but transgenic potatoes left the rats with intestinal reactions. At a minimum, Pusztai, argued, further study should be done.

Almost immediately, however, Dr. Pusztai was attacked and his results questioned. His institute was contacted by the companies that had paid for the research. This led to an investigation by his institute for what they called "unprofessional conduct" prompted by his having conducted a public interview (which he said was with permission from his institute) in which he disclosed his research results before they had been published. He also said some provocative and alarming things in the interview, including, "If I had the choice I would certainly not eat it," and "I find it's very unfair to use our fellow citizens as guinea pigs" (Randerson 2008). He was subjected to a formal ethical inquiry, suspended from his job, and banned from speaking publicly about his work (Robin 2010). The head of the Rowett Institute then used misconduct proceedings to seize his data. An audit criticizing his work was issued by the institute and approved by six reviewers. Pusztai issued his own rebuttal to the audit and received support for both his rebuttal and his results from another twenty-two European scientists.

Pusztai eventually published his report, not as an original research article but as a letter to the editor in the *Lancet*, after the letter was vetted by six additional reviewers and the editorial staff of the journal (Horton 1999). Two of these reviewers offered criticisms of his work (one thought it was a waste of time to review it, though it is not clear why). Other reviewers raised questions that were apparently sufficiently answered by Pusztai before publication (Ewen and Pusztai 1999; Rhodes 1999). In the end, the main scientific criticism of Pusztai's work was his generalizing of claims derived from a small sample size of rats. But the main reason most of his supporters think he was censured was that he went public with his findings that GE foods were potentially unsafe—a claim that industry supporters and many scientists sympathetic to industry disagreed with based on other research that showed no ill effects of GE foods in study animals.

In fact, the specific GE potatoes Pusztai studied never made it to market, although other GE potatoes were later developed and are commercially available today (including the Innate Potato, made by Simplot, designed to reduce blackspot bruising, browning, and the amino acid asparagine, a naturally occurring amino acid that can turn into carcinogenic acrylamide from frying). Despite the fact that his research findings may have been valued enough by the company to deter it from developing these specific potatoes (though we cannot know for sure), Pusztai's contract with the Rowett Institute was not renewed. His reputation continues to be questioned by media references to his ordeal as the "Pusztai Affair"; he is either castigated as a scientific fraud (among GE food supporters) or lauded as a hero (among anti-GE food activists). His fate and the degree to which he was made to suffer for his actions offer insight about the reasons for a broad anti-GMO movement in the United Kingdom and Europe—a movement that has had more impact and sustained visibility there than in the United States.

Pusztai's story offers important insights about how facts about harm have been produced over the past nearly thirty years (the same years in which glyphosate has risen to glory, or infamy, depending on which side you take) to form a counterfactual archive (Rowell 2003). To date, only a few other researchers have done studies that show harmful effects on the intestinal systems of pigs from GE foods and pesticides (Carman 2013). Before getting to that, it is also worth spending some more time on Séralini, who features prominently in the NASEM report.

Gilles-Eric Séralini is a molecular biologist based in France who published a series of articles on the possible health harms from GE foods. The contestation over his work is also referred to as an *affair*, in the most scandalous sense of the term. Séralini's early work evaluated the studies done by Monsanto that were used to establish safety claims for their Bt-modified corn and suggested that Monsanto's own data indicated hepatorenal (liver and kidney) toxicity in rats who were fed this transgenic food. That Monsanto researchers considered their own findings to be "biologically insignificant" was, for Séralini, a misreading and downplaying of the results (Séralini et al. 2007). This 2007 publication may have initially marked him as a researcher critical of GE foods, but it was not what made him famous; that was a later publication.

In his later work, he conducted experiments on the effects of GE foods and Roundup, modeling his research on studies that had been done by Monsanto scientists. Over the course of two years, he exposed rats to (1) a Roundup-tolerant genetically modified maize cultivated with Roundup (constituting at least 11 percent of the rats' diet), (2) the same maize cultivated without Roundup, and (3) Roundup alone (from 0.1 ppb in water) (Séralini 2012 et al.). That is, rather than studying the impacts of specific traits of GE crops as stand-alone vectors over a relatively short period of time (ninety days), he studied the health effects of all three kinds of exposure over a two-year window. In other words, he studied the kind of exposure that humans consuming glyphosate-rich foods might have. The results of his research suggested not only kidney deficiencies but also increased tumorigenesis in rats fed all of these combinations, especially in those fed maize grown with Roundup and those fed diluted amounts of Roundup in water. The effects were more pronounced in female rats.[2] Séralini interpreted the higher number of tumors in the treatment groups as evidence of low-dose tumorigenesis from all three exposures, with the last group (fed diluted Roundup in water) showing the highest number of tumors along with other signs of injury in the kidneys.

90 After using an unusually cautious pathway to publishing his results in the journal *Food and Chemical Toxicology* that included asking journalists to sign confidentiality agreements for advance access to the article, his findings were published. Immediately thereafter, criticism of his work began. Some

critiques focused on his use of tumorigenic rats (called Sprague Dawley rats) that made his results indecipherable: since tumorigenic rats are designed to grow tumors in a short period of time, any study of this length of time (two years) with these rats would show increased tumorigenesis. Criticism also focused on what some considered an inadequate sample size, especially since many rats died (though not necessarily from the exposures) in the two-year test period. His unusual commitments to "protecting" his research results from anticipated critique by having journalists sign confidentiality agreements led some to suggest this also rendered his work suspicious.

Séralini's research was also published following peer review, meaning at least these reviewers thought his work was accurate, reliable, and worth publishing, but extreme pressure was put on the editor and the editorial board of the *Lancet*, resulting in the journal's request that he withdraw his article (Fagan et al. 2015). (Critics suggest the pressure came from industry representatives or scientists paid by industry who suddenly appeared on the editorial board.) Refusing to do so, Séralini waited while the journal forcibly retracted his article. Thus, the Séralini affair had begun.

Séralini responded to the retraction in a publication, defending his methods and results, noting that he used the same laboratory protocols and study animals that were used in industry research. Criticism of his defense followed soon after from industry and industry-funded scientists, which was then followed by criticism of his critics, and then followed by published accounts attesting to evidence that the editorial board of the journal had been reconstituted to represent the industry's (i.e. Monsanto's) interests (Séralini et al. 2014b). Considering that the conclusions of Séralini's work were quite modest—he wrote, "Our findings imply that long-term (two-year) feeding trials need to be conducted to thoroughly evaluate the safety of GM foods and pesticides in their full commercial formulations"—the attacks on him and demands for retraction seem overplayed (Séralini et al. 2014a). Nevertheless, the juries of public and scientific opinion are wildly divided on his contributions to this day.

Séralini eventually republished his work in 2014 in the journal *Environmental Sciences Europe* (Séralini et al. 2014b). However, the controversy led to his being asked to leave his university, forcing him to set up an independent research institute where he continues to conduct research that shows negative health effects of GE foods and their associated pesticides in research animals. He even redid many of his own experiments, all showing deleterious effects from GE foods and Roundup in study animals (Séralini et al. 2007, 2012, 2014a, 2014b).

91

The Pusztai affair and the Séralini affair, among others, take up a great deal of airspace in the debate on GE foods. For some, the Séralini affair continues to provide prima facie evidence that science can effectively police itself in order to produce reliable truths. For others, these "affairs" are evidence of how science critical of industry is vulnerable to the most insidious censorship and how industry-driven science tends toward corruption, whitewashing, and outright lying about the facts.

Séralini's work was followed by that of others who came from his laboratory, including the geneticist Michael Antoniou at King's College London. Antoniou's lab has conducted studies that show even small amounts of Roundup result in higher rates of kidney disease and nonalcoholic fatty liver diseases in study animals, among other things (Mesnage et al. 2013, 2015, 2016, 2017). Based on this research, and taking up the concerns of a growing constituency of activists in Europe, the United Kingdom, and the United States, Dr. Antoniou joined forces with John Fagan, a molecular biologist, and Claire Robinson, a journalist-philosopher, to produce a detailed inventory of available scientific evidence to explain and counter the claims that GE foods and pesticides are safe in the book *GMO Myths and Truths*, which has been revised and updated several times (Robinson et al. 2015). Their work spawned and their results have been bolstered by a growing body of research investigating harm from GE foods with the combination of Bt and Roundup Ready traits as well as Roundup itself.[3] Antoniou became a vocal advocate for, in his words, using "science" to decipher claims about the facts on both sides of the debate over glyphosate's toxicity. Thus, in addition to offering publications that clearly show that Roundup is dangerous to the health of study animals, he has also published articles criticizing researchers who use faulty reasoning and evidence to claim glyphosate's heterogeneous toxicity (Antoniou et al. 2019). This includes debunking the ideas that glyphosate is a potential cause of autism (among other things) and that it can replace the glycine found in humans (Mesnage and Antoniou 2017). When I met him in 2016 at a GMO science retreat, he was committed to the idea that, in the end, good science would show that GE foods, and particularly glyphosate, were harmful to the health of humans. Still, many of his publications have been criticized by pro-industry reporters, who lump his work together with that of Séralini (Stiles 2017).

Most scientists who raise health safety concerns over GE foods and glyphosate have been subjected to a great deal of criticism, scientific discrediting, and character assassination, just like Pusztai and Séralini. Sometimes the attacks come from other scientists, and sometimes the criticisms are orchestrated by laypersons who generate blogs and websites in support of GE foods. The criticism often starts with a form of subtle harassment using FOIA requests for all emails sent on university or public servers. Paul Mills, an MD-PhD researcher at the University of California, San Diego, published the results of a study showing thousand-fold increases in levels of glyphosate in urine samples of Californians that were obtained and stored between 1993 and 2016 (Mills et al. 2017). Within months after publishing this information, he was subjected to invasive FOIA requests for his entire backlog of personal and professional email communications—a request his university told him they could not protect him from. Attacks like this usually turn quickly from FOIA requests to smear campaigns, tactics that are used especially on nonscientist public activists by talking about personal details of their lives that are meant to be discrediting.[4]

Criticism of those who question the safety of GE foods and glyphosate is so stunning that it sometimes backfires, casting doubt about the credibility and rationale of the accusers and defenders of industry. Even a quick perusal through the most popular websites (US Right to Know, GMO Science, GM Watch, and Biotech Salon on one side; The Genetic Literacy Project, GMO Answers, and Alliance for Science on the other) reveal a level of vituperation and hostility replicated in few other places where science is contested. A good example of this is the work of Jon Entine, whose expertise as a journalist got him appointed at the California Institute of Food and Agriculture Research at the University of California, Davis. He is well known for receiving funding from agroindustry and food corporations through this center. On his website, Entine claims to offer the only "real science" on these foods, quoting people he calls "real scientists" who agree with his views, and he persistently ties his defense of these food technologies to larger claims about the defense of science, *tout court*. In other words, anyone who says GE foods and pesticides are not safe is misreading the science, is anti-science, or is simply a fame-mongering activist with no scientific credibility.

To read these information streams or to watch the media they produce, such as the 2018 documentary/propaganda film produced by industry, *Food Evolution*, one might think that the only scientific work on GE foods consistently affirms them as safe (which is not entirely true). While it is true that the scientific consensus has been formed around this view, this fact is usually coupled with a portrayal that suggests the only people opposing GE foods are crackpot activists without scientific credentials or scientists who are easy to discredit (which again is not entirely true). Stone (2015, 2018) offers a very compelling account of the exaggeration and credibility debates on both sides of the GE food controversy.

93

Perhaps the best example of overreach and backfire in relation to industry-funded attacks is the case of Andras Carrasco, an Argentinian molecular biologist who, in 2013, found that exposure to glyphosate even in small amounts produced teratogenic birth defects in the embryos of frogs and chicks (Paganelli et al 2010). Carrasco began to lecture and publish on the results of his studies, but his work was soon made to seem controversial. His supporters suggest that because his research results would have a devastating impact on the massive GE soybean industry in Argentina, a full-scale defamation campaign against him was launched by his university, prompted by pressure from agrochemical industry and government representatives. Whether or not it is true that agrochemical industry forces were behind it, his laboratory was apparently stormed by lawyers for the consortia of agribusiness companies. Even the US embassy in Buenos Aires became involved in the investigation and punitive efforts that followed, apparently demanding that he be fired. Soon after this series of events, Carrasco was denied promotion at his university and died of a heart attack a year later.

Carrasco's contributions live on after his death in a kind of hero worship in testimonials like the following, by fellow Argentinians engaged in the battle for food safety: "From a little known scientist, in [a] few years he became a public figure and, for some, a hero. Last June, the School of Medicine of the University of Rosario (Argentina's third biggest city) established the 16th June—Carrasco's birthday—as the Día de la Ciencia Digna (Day of Dignity in Science), to celebrate the role of knowledge and of scientists in the service of the community (and not in the service of profit). Other Argentine universities already agreed on the commemoration day" (Adamovski 2014).

Even moderate critics of GE food technologies have often been unable to find neutral ground from which to question the facts about safety of these foods. Consider Belinda Martineau, the biochemist who helped develop the first GE food, Calgene's Flavr Savr tomato, who wrote a book about her concerns over the slipshod studies that were used to claim these tomatoes were safe for human consumption (Martineau 2001). She described her painstaking efforts to bring relevant data to the attention of her laboratory heads and to request that her information be included in evaluation processes at the FDA, only to be rebuffed and ignored. Her career as a scientist was severely curtailed by the backlash from her scientific colleagues.

94   Similarly, Dr. Judy Carman, the Australian biomedical researcher I mentioned above, described being sidelined and attacked for publishing studies that suggested possible harm from GE feed in pigs. Her studies, like Pusztai's, showed severe stomach inflammation as well as reproductive anomalies in live-

stock pigs and laboratory rats fed Bt and glyphosate-rich GE foods (Carman et al. 2013; Zdziarski et al. 2018).[5] When I contacted her in 2017, Dr. Carman told me that she had survived five attempts to sack her from her research institutes. At one point she hired a defamation lawyer who, she said, "helped a little." She noted that her work is regularly deleted from her website by what she assumes are "industry hackers." Carman's work, as I mentioned, is also regularly criticized by pro–GE food scientists such as Alison Van Eenennaam, a PhD and specialist in animal genomics and biotechnology at UC Davis. Van Eenennaam publishes epidemiological studies on bovine livestock fed GE feed (comparing livestock fed three things: GE foods, GE foods and pesticides, and non-GE food not treated with pesticides) that consistently show no differences in effects on livestock health (Van Eenennaam et al. 2007; Van Eenennam and Young 2014; Conroy 2018).

What is interesting about the scientific archive on GE food and pesticide safety is, as Glenn Stone (2015) has argued, how insistent everyone is about this being a zero-sum issue. The argument taken up by scientists on both sides of the debate is that only one set of facts can be right. Few approach the subject with the possibility that researchers producing opposing results about the safety and harms of GE foods and pesticides could both be right in their own ways. The positionality of those involved augments the degree to which the facts are made to seem reliable. But competing claims are often a result of noncomparable research designs, like comparing apples to oranges. Dr. Carman, for instance, looked at pigs' stomachs, but Dr. Van Eenennaam uses large epidemiological studies of growth rates for livestock reared for the market. One study used biopsies of gut tissues and the other used weight measurements taken on living livestock. The idea that pigs' guts (which are a lot like humans') are not at all like cows' guts (which are not like humans') is never mentioned, but would not be in any case since Dr. Van Eenennaam is not apparently interested in that kind of assessment of gut health. Nevertheless, both scientists use their studies to discredit the other.

When faced with the proposition that there are studies that show GE foods and pesticides are safe as well as studies that show they are not, few would read this as a situation of "both/and" in the sense that both could be correct (although there are some exceptions; see Benbrook 2019). Rather, the results are usually orchestrated as an "either/or" scenario in which the science that favors industry claims of safety prevails for the entire gamut of GE foods and pesticides, or the science casting doubt on all of this does. We might read these stories as forming a compelling reverse inertia in the swirl in which scale-up is not so much disrupted as derailed. Like a new point of

95

departure for a clustering assemblage of scientists, research findings, and experimentally produced facts, these counterfactuals derail the persistent demand for a singular consensus that is stable and, as the NASEM scientists and their spokespeople often say, "is settled once and for all."

My goals in this chapter have been to introduce you to the trouble glyphosate has been causing in the world of science and to impart to you a sense that glyphosate—the chemical that comes along with almost all GE foods—is at the center of a prodigious controversy over the utility of the scientific consensus. The consensus is an unstable entity. But this does not prevent it from being used to make claims about how we should feel about the toxicity of glyphosate and GE foods. Indeed, the consensus becomes an arbiter of not just the facts about this chemical; it is also used to arbitrate new kinds of debates over the trustworthiness of science itself. In this sense, glyphosate disrupts a conventional reading of both how chemical harm is managed by industry and how scientific consensuses are diffracted into multiple competing resting points. We do not get a consensus with glyphosate, but rather many competing consensuses. As glyphosate gets taken up and deliberated by other constituencies, it is given new kinds of powers that lead to yet more different competing consensuses. I turn to this in more detail in the next chapter.

Consensuses, Academic Capitalism & the Swirl

According to survey reports, roughly 88 percent of scientists believe that GE foods and their associated pesticides are entirely safe (forming a consensus view), while under 40 percent of the lay public believes this view, meaning more than 60 percent of the public does not believe the scientific consensus (Plumer 2015). One might be inclined to think that the lay public's wrongheadedness and mistrust of the scientific consensus stem from the fact that most of the public is scientifically illiterate; the data are therefore an example of collective public ignorance. The magazine *Scientific American* weighs in with this view, discrediting all critics of GE foods and pesticides by suggesting they suffer from a form of psychological delusion. In an article entitled "Why People Oppose GMOs Even Though Science Says They Are Safe," Blancke (2015) argues that GMO criticism is an example of what psychiatry calls the Dunning-Kruger effect, in which "the less you know, the less able you are to recognize how little you know, so the less likely you are to recognize your errors and shortcomings" (Pryor 2018).[1]

Indeed, many scholars have argued that the problem in the debate over GE foods is one of insufficient communication of the scientific facts to the lay public (Ladrum et al. 2019). Although it is easy to agree with this explanation, especially when GE food skeptics are lumped together with climate change deniers, anti-vaxxers, or other contemporary conspiracy theorists, I do not think it can entirely explain the problem at hand. In fact, I would argue that what you believe about the safety of these things depends on where exactly you are situated in the swirl. To make this clearer, I explore how the debate over glyphosate and GE foods raises questions about what is meant by the scientific consensus and how it has taken particular shape in relation to these chemical-rich foods at this particular time in the history of late neoliberal capitalist science and information streams. In particular, I want to go further into the problem raised by the idea that we could explain the conflict in scientific claims about GE foods and glyphosate by blaming either distortions produced by industry funding, on one hand, or on the collective ignorance of the public (and a growing group of scientists), on the other. The scientific consensus, I would argue, has to be parsed as a peculiar object and accomplishment of science that may have outlived its utility in an era where it is no longer possible to claim either that industry investment in science leads to untrustworthy facts in relation to chemical harm or that the dissent to consensus views is a product of scientific ignorance and conspiracy thinking.

Naomi Oreskes and Erik Conway (2011), in their book *Merchants of Doubt*, take up the argument that despite its persistently and definitionally contested status as an artifact of the scientific process, the scientific consensus can serve as an arbiter in most situations where scientific facts are being contested, noting that it is a particularly reliable compass in navigating through sciences that have been corrupted by what they call *industry bias*. Their case in point is the linked histories of the tobacco industry and climate change denialism. They show how industry-funded scientists "used their scientific credentials to present themselves as authorities, and . . . used their authority to try to discredit any science they didn't like" (Oreskes and Conway 2011, 8). Some of the scientists who defended the tobacco industry in lawsuits were the same ones who defended oil companies against charges of destroying the planet through climate change. In most cases, these scientists' industry funding was for producing scientific facts that would not *displace* consensus facts but simply produce *doubt* about them.

In the case of GE foods and glyphosate, the scientific consensus has been from the start almost entirely aligned with industry priorities. In fact, one of the most vocal critics of the scientific activism against GE foods and their pesticides is Dr. Henry J. Miller of the Hoover Institute, who authored a 2018 Newsweek article titled "The Campaign for Organic Food Is a Deceitful, Expensive Scam" in which he suggested this activism is funded by the organic food industry. Interestingly, Miller was one of the merchants of doubt decried by Oreskes and Conway for peddling falsities in the 1980s to create doubt about the scientific consensus over the safety of tobacco. Nowhere in his writing against GE food activism does Miller disclose his own funding from agrochemical industry sources, which includes having been paid by Monsanto to ghostwrite industry reports attesting to the safety of GE foods (Malkan 2018). Thus, while Miller was once paid by the tobacco industry to produce doubt, he is now paid by the agrochemical industry to attack those who cast doubt on the scientific consensus that favors industry. Perhaps once an industry shill, always an industry shill. But is the fact that industry has invested the most time, energy, and money into producing the consensus on GE foods enough of a reason to dismiss the consensus view that such foods are safe?

I mentioned that although Monsanto perfected the marriage of industry and academic research in its biotechnology pursuits in the 1970s, the idea of industry-funded science was not new. For-profit drug development in medical professions was alive and well prior to World War II (Gabriel 2016; Kaufman 2015; Starr 1984), eventually prompting skepticism about the biased nature of for-profit drugs peddled by clinicians of various skills and training. The Flexner Report brought more scrutiny of these practices, and efforts to crack down on drug sales by clinicians emerged alongside efforts to clean up and standardize medical education over the decades after. By World War II, government investment in academic science spurred the growth of science research, making it unnecessary for research scientists (particularly medical researchers) to rely on industry funding. The growth of publicly funded academic sciences helped popularize the idea that academic research could be trusted in ways that industry-funded research could not (Gaudillière 2017; Bero 2018; Chiu et al. 2017) and also propelled many of the concerns that researchers at Monsanto confronted when they tried to augment such collaborations, as Schneiderman's work showed. Nevertheless, by the 1980s and the passage of the Bayh-Dole Act, which made it possible for universities to obtain patents on federally funded research, academic institutions in the United States opened the floodgates for industry collaborations (Schurman and Munro 2010, 13).

Academic capitalism is normative across most scientific fields today and is often, though not uniformly, presented as a neoliberal success. Slaughter and Leslie (1997) argue that to frame the problem of industry funding in terms of corrupt versus trustworthy science is to miss the way market logics slide in and take over the entire edifice of knowledge production, blurring profit-driven goals with those that may not be profitable but are nevertheless examples of reliable science. This is part of the logic that Schneiderman used to convince his academic colleagues to come on board with Monsanto

99

collaborations (the other part being that industry could actually fund academic research better than government). The subtle impacts of industry funding are not debatable at the level of validity or reliability but rather, as Isabelle Stengers (2018) notes, at the level of more basic questions of what can and cannot be legitimately studied. She calls industry-driven science "fast science" in that it persistently turns questions of scientific reliability into questions of application.[2] The drift over the twentieth century, she notes, was toward one side of that equation in which the very prospect of abstract thought, uncertainty, and messiness in research was increasingly displaced by the demands of narrow and siloed fields of study that are orchestrated by the efficiencies and focus of industrial applications. The very notion of reliability has shifted toward the value regimes of industry (Stengers 2018, 8).

This digression into the debate over the perils and profits of academic capitalism may feel like a departure from our narrative about glyphosate, genetically modified crops, and consensus. But Stengers's case in point in her essay on fast science is the sacking of the academic agroecologist Barbara Van Dyck. Van Dyck was fired by her university for having publicly participated in activism to stop an experiment on genetically modified potatoes in Wetteren, Belgium, that she was involved in. The university sacked her on grounds that she was not doing her job. Stengers's point is that to sack a researcher for acting publicly in relation to the knowledge production she was involved in is to unveil the problem of fast science and the narrowing of what counts as "science" at all under the burdens of industry goals.

Market-driven scientific rationales have, of course, been at work in the creation of GE foods and pesticides since their inception, and such rationales appear at every stopping point of regulation and evaluation, from their discovery in the lab to their distribution and perfusion through the entire US food supply. Again, this fact does not make the foods or pesticides harmful, nor does it necessarily imply that the science that created them was not good science or that the scientific studies of their safety were necessarily corrupt. The creation of these foods and the claims about them and their pesticides are products of the commercial goals of the companies that made them. It should not be a surprise that industry scientists did not look for, or find, reasons to keep these products off the market, as Joe Dumit (2012) notes for the pharmaceutical industry. Indeed, they are simply doing what good capitalists do. What is perhaps more surprising is how hard corporations, including Monsanto, have worked to keep information that might do damage to these commercial goals from seeing the light of day.

It has been documented that agrochemical industries have used ethically questionable tactics to subvert the normal peer review and meritocratic processes of science: paying for attacks on scientists whose work casts doubt on the safety of these foods and pesticides, paying "unaffiliated" scientists to ghostwrite articles in favor of their safety, and paying pro-industry scientists to serve on editorial boards, scientific review boards, and peer review boards to ensure that publications about GE food technologies are positive and that articles producing doubt about their safety are retracted or never published (Gilham 2017; McHenry 2018). Industry defenders have argued that they have had to use these tactics to get fair treatment in the face of great and misguided public suspicion and activism against these foods and chemicals, not to bias the science. Critics of the agricultural industry argue that these activities point to the corruption of the science on which the consensus has been built. The swirling begins.

Efforts to discredit the scientific consensus on the basis of it being untrustworthy because of its ties to industry are compelling, but impotent. We have long passed the turning point for being able to claim that industry funding in and of itself corrupts science, even when we can show that there have been systematic attempts to silence the voices of scientists who argue with a consensus view or present counterfactual scientific conclusions. Similarly, it would be shortsighted, and perhaps incorrect, to argue that, because we can document industry involvement in the scientific policing of facts about GE foods in ways that consistently favor industry goals, this means the counterfactual science is the only reliable science.

This is the problem. The sleight of hand made possible in academic capitalism is that it blurs the lines between *reliable* and *biased* science such that the very conditions for evaluating scientific reliability make us incapable of separating the needs and goals of industry from the built-in biases that these needs produce in the research itself. These conditions have produced a scientific "zone of indistinction" over things like safety or the deciphering of whose and which scientific results should be trusted when it comes to possible harms or risks (Kaufman 2019). In the case of GE food safety, this indistinction makes it impossible to either discredit or believe the scientific consensus.

Despite the proliferation of evidence from laboratory science and from the social histories of censorship and censure of those who have claimed that GE foods and glyphosate are potentially harmful, claims to a scientific consensus that says they are safe persist. The discrediting seen in relation to counterfactual science can explain this to some extent, but so can the science upon which the consensus is based. That is, there is plenty of scientific evidence that suggests these foods and their companion, glyphosate, are safe. What spending time on the counterfactual record helps us understand is that the deliberative field in which glyphosate and its GE food companions live has become unmanageable in its multiplicity. The idea of a singular consensus is made to seem irrelevant. As glyphosate travels by way of plants into foods; into and out of research laboratories and pigs, mice, and rats who are forced to ingest it; and even onto the pages of reputable scientific and medical journals, glyphosate has been asked to carry very heavy loads. In each case, glyphosate plays a role in producing facts about its safety that lead in different and competing directions. The scientific consensus has also been asked to carry heavy loads—perhaps a weight it can no longer carry.

Glenn Stone (2015) calls the GE food debate a good example of Gregory Bateson's description of *schismogenesis*, in which most of what gets published and publicly pronounced gathers its momentum and credibility from the absence of a reasonable and neutral conversation coupled with a polarization of the possible positions one could take and the science one can read. This is not just a problem for the layperson; it is a problem for scientists themselves. The deep grooves of advocacy both for and against GE foods and their pesticides have been carved by scientists, industry workers, policy makers, and activists who always end up on one side of the debate or the other. There is, as Glenn Stone (2017) notes, no neutral ground to stand on. Almost everyone writing about the safety or potential harm of these objects is quickly attacked (and their work questioned) by the opposing side.

Stone calls for more "honest brokers" in the effort to decipher harms from GE foods and pesticides, holding out for unbiased facts that could float to the surface and inform us about how safe such foods really are. I would argue, however, that what we are witnessing in the debates over GE foods and pesticide safety does not reveal a problem of dishonesty (although there is evidence of that, too) so much as a problem of the production of competing certainty that was built into glyphosate from the get-go, and from the food-cum-chemical objects that are neoliberal, late industrial food. By following glyphosate in and through food systems, environmental and toxicology studies, clinical offices, sick bodies, and even peer-reviewed publications, we gain some sense of how its multiplicity and its multiple potencies make it impossible for facts to settle. Attempts to form consensus have not settled the facts. Instead they have proliferated and produced many consensuses.

What I am pointing to here is both like and more than the usual story about how normative science produces—in fact thrives on—uncertainty, or

102

what Proctor and Schiebinger (2008) call *agnotology* (see also Boudia and Jas 2015). To be sure, the scientists who have come into glyphosate's sphere have forged ahead because they have not been entirely sure what they would find when they fed the chemical to rats or pigs or mice. But glyphosate has also asked more of these scientists: that they forge from this work a stable consensus. The demand for a consensus has been so great that it has caused a lot of people to attack their fellow scientists and make them lose their jobs.[3]

My point in bringing you into this mire of competing certainties about glyphosate and its companion foods is to suggest that glyphosate offers a bird's-eye view of how chemicals that may or may not be harmful are able to simultaneously inhabit contradictory truth spaces. The science about chemical harm in relation to glyphosate remains problematic in its multiplicity. Glyphosate thus weaves together an interesting swirl of normative science, providing us with a great example of how chemicals can disrupt not just cells, microbes, and genes, but also scientific efforts to establish facts by persistently arbitrating them and refusing to let them settle.

Susan Leigh Star and James Greisemer (1989) might have called glyphosate a boundary object in the sense that it has the capacity to thread together disparate fields of study and through lines of evidence and analysis simply by appearing in multiple studies as the same chemical. Glyphosate helped the NASEM metastudy authors scale up their claims about its safety without discussing the things that are made obtuse and obscure by the extraction and boundary object instrumentalities that such a study relies on. It works at the boundaries between different kinds of science and evidence, making various experts with nonshared knowledge feel that they can agree on some things.

And yet, even as a boundary object, glyphosate has refused to offer the kind of stability that these augmentations and collaborations strive for. In the chapter before this, I noted that there is a large and growing archive from various scientific disciplines that points to possible harms from glyphosate. Some of these were not available at the time the NASEM metastudy was done, while others seem to have been overlooked or were presented as having results to doubt. In the same year as the NASEM report was published, the environmental science scholar J. P. Myers and several colleagues from medicine and biological sciences conducted their own metastudy (using the same shingling and scale-up techniques as the NASEM study) and published it in the journal *Environmental Health*. The article, titled "Concerns over Use of Glyphosate-Based Herbicides and Risks Associated with Exposures: A Consensus Statement" (Myers et al. 2016), argued that the preponderance of evidence about glyphosate suggests that it indeed poses health risks to humans. Their study is based on eighty peer-reviewed scientific publications, including laboratory animal studies, most of which were available at the time of the NASEM study. Their conclusions are in direct opposition to the NASEM report. First they remind us that, contrary to what the NASEM study and industry maintain, bioaccumulation is a problem because glyphosate is rising in humans and human environments, saturating environments once thought to be capable of quickly biodegrading it.

> (1) GBHs [glyphosate-based herbicides] are the most heavily applied herbicides in the world and usage continues to rise; (2) Worldwide, GBHs often contaminate drinking water sources, precipitation, and air, especially in agricultural regions; (3) The half-life of glyphosate in water and soil is longer than previously recognized; (4) Glyphosate and its metabolites are widely present in the global soybean supply; (5) Human exposures to GBHs are rising; (6) Glyphosate is now authoritatively classified as a probable human carcinogen; (7) Regulatory estimates of tolerable daily intakes for glyphosate in the United States and European Union are based on outdated science. (Myers et al. 2016, 19)

They then turn to the medical evidence of harm in research animals. One particular concern is how glyphosate "provokes oxidative damage in rat liver and kidneys by disrupting mitochondrial metabolism at exposure levels currently considered safe and acceptable by regulatory agencies," drawing from numerous studies that were available before 2014 (Myers et al. 2016, section II). They note that the presence of glyphosate and its metabolite, AMPA, in both laboratory and farm animal liver and kidney test samples is ten- to one-hundred-fold higher than levels found in other bodily tissues of these animals, also citing Séralini's studies. They conclude with epidemiological studies showing that chronic kidney disease is higher in male agricultural workers than in other populations in the United States, that glyphosate disrupts endocrine signaling systems in wild and laboratory animals, and that it is associated with cancer and that the incidence of non-Hodgkin's lymphoma had doubled in the United States between 1975 and 2006. They conclude with a final prospect of harm, noting that glyphosate is a powerful chelating agent that can sequester micronutrient metals including zinc, cobalt, and manganese. Where was all of this information hiding when the NASEM study was being constructed?

Thus, glyphosate has disordered the scientific consensus like a gust of wind, blowing it from one point of certainty to another. While it has had a lively role in the scientific consensus that maintains GE foods and pesticides are safe, it has also had a lively role in the production of other consensus views, including one that says the opposite—that glyphosate (and therefore most GE foods) are not safe for humans. These are accompanied by others that have emerged in the public, legislative, and regulatory deliberations of it (as we will see in the next chapter). Glyphosate refuses to belong to any one consensus, leading to a veritable war over certainty in the sciences about this chemical (Hilbeck et al. 2015).

I have offered this churn through the scientific archive on GE foods and pesticides not just to argue that because the science is contested, consensus is itself a slippery affair. Scientific relations with glyphosate are fraught because of what glyphosate is and how we currently (in the United States at least) do science and form consensuses. As glyphosate moves through the worlds of scientific consensus making, it has proliferated competing facts about it, not unlike what happened to it as a chemical in Monsanto's laboratories some forty years earlier. What this means is that whatever you might want to say about a consensus on glyphosate, you are likely to be able to find other studies that point to the opposite and suggest an alternative consensus. Glyphosate has kept the consensus unstable and on the move, creating diffracted certainties based on multiple potencies that have enabled it to move in different consensus-making circles where it repeatedly drives forward claims to certainty. These consensuses provide diffuse and multiple set of certainties that are filled with hauntings, even if no absolutes, in relation to harm.

I think of the ability to read the available science about glyphosate as calling for specific kinds of titration—or rather, of joining the arbitrary and ever-changing flight patterns that emerge from the constellations of data, sentiments, politics, and experiences that come with living with glyphosate. This might be similar to how bodies titrate chemicals from ordinary living in the Anthropocene, sometimes clustering here, other times there, and settling unpredictably in tissues that can be damaged. I would argue that glyphosate helps me to think about scientific facts in the same way that I think of bodies as showing the heterogeneous effects of the chemicals in transit inside of them. The facts on their own do not produce clear and linear lines of singular certainty; they cannot sort themselves out and produce consensus (or uniform effects) alone. The facts are, even in their creation, already situated, relational, and distributed (Murphy 2017a), and their utilities are spread out along trajectories that sometimes lean toward what industry wants, sometimes toward what regulatory agencies want, and sometimes toward what different groups of scientists and activists want.

Facts do not make consensuses; people do, by holding some kinds of evidence close and others distant, by following the traces of harm to their resting points in organs, tissues, soils, urine samples, laboratory rats, and peer-reviewed journals. By the same token, glyphosate has played a starring role in both forging a scientific consensus and undermining it. The metastudies and practices of evidentiary scale-up that have been used to form the consensus that GE foods and glyphosate are safe is swept away by this chemical's emergent role in destroying that consensus, revealing *the swirl*.

The Swirl

Wading into the debate over the safety of glyphosate offers some sense of how facts about it settle solidly in certain places but generate other certainties that stand in direct opposition to the former when circulated to other places. The facts move around different scientific archives and shift relationships to the firmness of certainty about them, like chemicals in different parts of the body, causing one cellular irruption or interruption that cascades into others in ways that are not always reproducible or comparable once they leave the laboratory, the farm, the nation, the body. To apprehend this movement, consider relationships not just to chemicals but also to the heterogeneous archives of scientific knowledge about them. In doing so, the first thing that becomes apparent is that consensus gives way to the swirl.

What exactly is this swirl? Consider the swirling, undulating metaformation of data and things that resemble a clustering flight of birds (perhaps starlings) that fly in a tight formation, forming temporary yet solid-like cloud formations out of random bits of data in flight. The visual reference of the swirl in relation to GE food safety is a swarm of starlings as they cluster in the evening skies in patterns that signal affiliation and a managed intentional randomness. The swirl generates patterns of material interaction that remain alterable in flight, clustering information, evidence, reports, and bodies into data sets that appear tangible and almost solid, clustering in ways that momentarily give the appearance of fluid solidity. Fragmenting and forming multiple different formations of the swirl, then rejoining to form one, the objects in the swirl get swept up in movements that give them collective density. Swerving here and there with an inertia (*quorum sensing*, as it is sometimes called with birds), *clustering* conjures specific kinds of tangibility, visibility, and the appearance of temporary firmness.

The Italian physicist Giorgio Parisi calls this movement of starlings a *murmuration* (Cavagna et al. 2010)—"a shape-shifting cloud, a single being

moving and twisting in unpredictable formations in the sky . . . as if it were one swirling liquid mass" (Heimbuch 2019). Parisi figured out that these group movements are possible because of the way the sense perception of each bird is augmented beyond its individual capacity as it senses the birds in its immediate vicinity (up to seven birds or so). He and others consider this movement a protective phenomenon, enabling the birds to collectively seem larger than they are as individuals and thus avoid predation. Because this perceptual phenomenon happens no matter how many birds are in the group, Parisi calls it "scale-free" correlation. "Scale-free correlations provide each animal with an effective perception range much larger than the direct interindividual interaction range, thus enhancing global response to perturbations" (Cavagna et al. 2010, 118–65), though it is not fully understood what biological mechanisms or levers of collective movement are at work here. Every shift in perturbation is called a *critical transition*.

I think of the scale-free correlations of the swirl of glyphosate as working along the same lines, with movements that give individual points of evidence and chemical traces the feeling of largeness, and with critical transitions that make certainty unstable. Sometimes the flash points of certainty lean toward industry desires; other times they are pulled toward activist interests or counterfactual claims. Sometimes these larger-than-life movements describe glyphosate itself, with traces of visibility called biologically insignificant (or, sometimes, carcinogenic) in soils or in organs. The movements entail practices of scale-up but then become, in the murmuration of consensus building, scale-free, and thus capable of movement changes based on any number of clusters of conviction.

The swirl is not a disinterested movement (the elements of the swirl are partly directing themselves). Different kinds of "interestedness" force a directionality of movement and inertia in one way or another, contingent on the truth claims, technologies, and interests of the constituencies that the swirl lines up. The swirl is an alternative to the idea of a singular consensus, an ephemeral formation of certainty consisting of data, technology, publications, chemicals, and facts. The swirl directs our attention not to the knowledge debates that operate in the GE food world but rather to how certainty travels in and through things (chemicals, data, algorithms, and archives) in unison even while remaining unstable. The swirl of glyphosate gathers information
108 and material things to form transmuting architectures of certainty that are part of the human repertoire of experience and also push beyond knowledge, each shift being a kind of critical transition to new material possibilities and prospects. Settling at one moment on a consensus that shifts to another,

then another. Settling other times in cellular irruptions of genetic mutations, forming new microbial communities in guts, causing chronic dysbiosis, and sometimes passing through cells and bodies and showing up in urine
or breast milk. Just because it is in motion does not mean it does not change
things or leave lingering effects. In this sense, the swirl is useful for thinking
about not only scientific arbitrations but also clinical experiences and the
habitations of glyphosate in many different places.

The swirl brings particular clarity to the problem of contested science.
The metastudy is one way of managing the demand for and presence of
ever-more-certain kinds of evidence and is itself a good example of the swirl.
The technological machineries needed to manage these data now drive the
enterprises of science and determine the terms of deliberation over things
like the scientific consensus. Whereas it is commonly assumed that putting all the available data together will produce a reasoned, factual, and fair
way to engage in biopolitical management of things such as GE foods and
glyphosate, this assumption proves incorrect again and again. Within the
swirl, there is no resting point, no black box that permanently stabilizes
and closes off the facts upon which scientific work progresses, no sense of
progression arising from the assemblage of things and social networks that
allow us to move toward closure. What we have in the swirl is endless flight,
and yet this movement invites me to think about how things like chemical
harm also remain in motion—with glimpses of certainty here, irruptions of
uncertainty there, on bodies, in soils, in EPA panels.

I do not think of the swirl as displacing other conceptual work from the
many science studies scholars who have offered various ways of talking about
knowledge production and the tactics of producing stability in science. Thus
I take as axiomatic that some things are already given about the relationships
between scientific action and the formation of consensus as demonstrated by
many other social scientists. Glyphosate points to the nonlinear qualities of
such efforts, forming more of a spinning-looping-wallowing of achievement
rather than anything like progress. The swirl, with its movement and heterogeneity and temporary resting points and scale-free correlations, offers
some credence to the idea that science never settles but clusters and lumps
and loops in ways that give rise to forms of temporary or soft certainty and
that in any case leave traces of how things have changed. The swirl captures
something of the way that chemical harm works, normatively, ephemerally,   109
and in ways that settle on the prospect of temporary kinds of certainty only
to give way to new iterations of facts and actions around other provisional
certainties. Forms of life are not harmed in the same way by glyphosate as

it moves through bodies and permeates different tissues at different rates in different species, and the question of just how harmful it is to bodies remains contested despite there being concrete evidence of this possibility.

The ephemerality of the swirl offers a conceptual and material framework of knowledge, and it sheds insight on how things (like chemicals) work in relation to knowledge shaped by the materialities of the chemical itself, including the scientific laboratories, funding streams, activism, publicity, cells, DNA, and tumors that have coalesced into material things alongside the chemical as it has traveled. The swirl takes as given that science is driven by uncertainty in ways that must produce unknowns (Proctor and Schiebinger 2008), and that like normative practices in many sciences, particularly those that deal with chemical harm, scientific research thrives on instability. The swirl builds on the idea of the mix of human and material relationships that form assemblages of knowledge production and experimental environments in science (Latour 2007) and traces these networks in motion (Deleuze and Guattari 1987). The swirl also attends to the way advocacy and activism enter into the picture, themselves driven by the potencies of glyphosate in its many sites of impact in farms, laboratories, bodies. In this sense, the swirl offers an attempt to accommodate both the social constructionist view that scientific consensuses are always provisional (as are scientific facts) with the high-stakes effort to trace chemical harm and arbitrate harm even when there is no reliable consensus. The resting points of certainty and of harm reveal clusterings of things that can make evidence seem larger than it might otherwise be.

The swirl foregrounds the clustering of knowledge in relation to the expectations we hold for it to settle more permanently, to give traction that will alter the course of history, to heal a body, or to sway an EPA panel or a publicly constituted jury. Also, in pointing to the ways that the swirl is both about and more than knowledge production far beyond scientific venues, I think it helps solidify the movement and evasions of science that have emerged around chemical harm (Murphy 2006; Boudia and Jas 2014b), in which it swiftly becomes clear that the diverse constituencies involved in deliberating their presence in the world cannot come to agreement (Boudia and Jas 2014a). In some concrete ways, the swirl might help us think about how chemical swirls also configure what can and could be said about their pathways of injury. In this sense, I think the swirl builds on a shift not in relation to how knowledge works in the effort to trace chemical harm, but in how we read knowledge-producing activities in and through the life of chemicals in ways that enable us to trace their pluri-potentiality.

The formations of the swirl are made up of things relating to one another: studies, metastudies, statistical technologies, constituencies of the chemical, and the chemical itself, and its many potencies. Glyphosate's swirl includes the scientific archives that circulate to various scientific and lay communities (from bloggers to policy makers sitting on EPA panels) as well as the chemical that enables some facts about it (its inner truths) to be seen in some places and other facts to be seen in other places. Like the murmuration of starlings that coalesces in a uniform flight path, the various elements of the swirl are attuned to the way its other elements are pointing, the preponderance of evidence that starts to surface and create a kind of certainty about the harms of glyphosate and its safety, the onboarding of new interest groups and legislators and activists who have committed to counterfactuals; these collectively trigger critical transitions to make the swirl seem larger than any of its individual elements. The augmented perception of movement one way and not the other enables various constituencies to claim consensus only to be undermined by the swerve to a new direction, the pull of certainty that has found another path, the formation of multiple different swirls that veer in opposing directions. Like the quorum sensing that is speculated to be at work in the formation of murmurations, we might think of these coalescings as types of scale-free correlations that can grow vast and overpowering or remain small and dense yet still visible and impactful. As Tsing (2015) noted for the problem of scale in the architectures of plantation capitalism, I suggest that scale-free correlations work to counter the architectures of knowledge and exposure in relation to agrochemical harm, opening space for contemplating what the indecipherability of the facts can lead to in relation to accountability and activism.

The swirl—not as metaphor but as model—has particular saliency in our times, or perhaps under the burdens of the perceived scientific crises of our times, as the production of a scientific consensus about so many things seems to be so urgent and yet so persistently up for grabs under the pressures of scientific capitalism, industry complicity, regulatory capture, and the work of publics who craft platforms (perhaps plateaus) of certainty and activism to affect and activist politics (Boudia and Jas 2014b). Still, at a time when the idea that scientific archives can be called on to settle so many of our political, social, and material crises concerning chemical exposure, the persistence of a particular problem around chemicals like glyphosate continues to produce skies full of competing facts that all might be reliable. When the overproduction of facts calls forth the need for metastudies and big data technologies, the swirl arises as a refusal of singular claims to truth

Consensuses & Academic Capitalism

and to the prospect of consensus. Said another way, from the perspective of glyphosate, the formations of knowledge, technologies, archives, sites of action, and deliberations looks more like a swirl than anything else. The swirl's potency lies in these movements of clustering, of hovering together to make certainty seem larger than any of its individual facts, sequentially displacing and reforming as different constituencies grasp hold of different qualities of the chemical and the chemical lets them claim one thing or another about it. And that is exactly how glyphosate works as it travels to so many places. In Murphy's (2017a) sense, we might consider the swirl an instrument of distributive activist justice.

I want to emphasize that the swirl is more than a metaphor for knowledge. It is a model for capturing the movement of glyphosate in its penetrations and shaping of the worlds it inhabits. To be even more concrete about this idea, one might consider the impact glyphosate is probably having on birds like starlings and on their patterns of murmuration. Over the past decade, the world has seen multiple instances of mass starling death in which the murmuration brings them too close to the ground and they crash to their demise (Touray 2019; Dawson 2021). Although uncertainty persists around the cause of these mass starling murmuration disasters, arguments have been made for their being tied to chemical exposures (by proximity to chemical plant leaks, for instance). I would also suggest a consideration of the evidence of glyphosate's impact as an endocrine-signaling disruptor chemical in birds (Markman et al. 2011), and then speculate (since the research is not available) about how this disruption might, as a chemical impact, affect the flight and scale-free correlating abilities of birds. To come full circle, we might consider the off-balance disruptions of murmuration and possibly the end points of bird deaths by crashing into the ground in relation to the way that the swirl constituted by the presence of glyphosate has its own deaths and morbidities in both human and animal populations to account for.

Glyphosate scatters and diffracts its impacts as it moves through the world, and along the way it also diffracts the science about it. In this the swirl produced by glyphosate is both in the knowledge systems about it and in the worlds it travels, swirling in and through atmospheres where it is sprayed, penetrating soils and water and foods, whence it is transported into bodies where it may kill microbes and cause cancer, among other things. It may even harm starlings who inhale it on their evening rounds in agricultural territories. It would be a mistake, however, to claim that the operations of glyphosate and the swirl are chaotic, even when they are disrupted from the expected paths of flight. The operations

112

of the swirl form beautiful and horrific formations that can be mobilized for deliberative action. The swirl enables certain knowledge practices and claims to coexist, generating capacities for different constituencies over time and space to use these resting points for multiple kinds of distributive justice (Murphy 2017a). In this capacity, the swirl enables us to consider both how researchers and regulators and chemical industries fuel consensus practices about chemicals and how chemicals themselves take on more-than-human capacities for care in legislative action even as they cluster in cancerous and disrupted cellular formations, as we will see next.

7

We want to be clear.... All labeled uses of glyphosate are safe for human health and supported by one of the most extensive worldwide human health databases ever compiled on an agricultural product. In fact, every glyphosate-based herbicide on the market meets the rigorous standards set by regulatory and health authorities to protect human health.
—Official Monsanto statement in response to IARC announcement that glyphosate is a probable carcinogen (2015)

Glyphosate Becomes an Activist

After many years of deliberating and changing conclusions based on animal studies that were primarily coming out of Europe and the United Kingdom, in 2015 scientists working with the International Agency for Research on Cancer (IARC), headquartered in Lyon, France, gathered up more of the scientific literature on glyphosate and formed a consensus determination that the chemical was likely carcinogenic to humans (IARC 2015). Since the IARC is a semiautonomous unit of the World Health Organization, the latter took up this finding and classified glyphosate as a category 2a carcinogen, meaning there was limited evidence of its carcinogenicity in humans but reliable evidence of such in laboratory animals (Guyton et al. 2015). The arbitration process focused on one possible risk: glyphosate's ability to cause non-Hodgkin's lymphoma. The swirl, ever on the move, was morphing toward another new resting point around which to orchestrate another consensus in a sea of other consensuses.

The effects of the IARC report and the WHO classification were swiftly felt, and calls to ban glyphosate-based herbicides across the European Union were soon made (Kelland 2017; Arcuri and Hendlin 2019, 2020). Almost immediately, lawsuits against Monsanto were filed in California by people with non-Hodgkin's lymphoma who had been contacted by anti-GMO activist groups who believed these cancers were caused by long-term exposure to Roundup.

Recall that since the science about glyphosate's harm is as fraught as the science of GE foods' harm in general, glyphosate activists must contend with the same sense of contested certainty as the rest of the GE food activist community. Because glyphosate is an ontologically multiple thing with the capacity to be potent in many ecospheres, including those that have aroused political action, it has become (in its own agentive way) what might be called a *pluripotent activist*.

As expected, soon after the classification of glyphosate as a probable carcinogen by the WHO and the IARC, this finding was immediately contested by scientists at the European Food and Safety Agency (EFSA) who revisited the original data in the studies used by the IARC and argued that details were overlooked, evidence was not read correctly, and the conclusions were unreliable (Portier et al. 2016). One line of questioning referred to other evidence that suggested there was no harm to humans exposed to acceptable daily intake levels (Niemann et al. 2015). The more critical line of skepticism concerned whether the comparisons of the presence or absence of tumors between test and control groups in the studies that were being used (the same approaches used by the authors of the NASEM report and the EFSA) were sufficient to establish carcinogenicity. The critics specifically focused not on the graphs of processed data but on the visible surface volume of nontumorous tissue that was seen in the raw data: the images and classifications of cells from animals that had been exposed to glyphosate. Based on their reading of the micrograph images, they claimed that the original researchers underestimated that area, thus making carcinogenicity uncertain. Using technologies available to read things like this, to make clear what is tumor or not tumor on a square millimeter of tissue depicted in images of slides in publications, the IARC findings were made to seem questionable enough to cast doubt on the new consensus that was forming around glyphosate being unsafe.

In turn, protestations against the IARC ruling on grounds that the science was incorrect were then contested by IARC scientists and supporters who said that the questioning of the IARC report by the EFSA was itself further evidence of an industry-backed attempt at coverup and whitewashing (Wild 2018). Reuters reporters got involved in the sleuthing over the ruling as well. An investigative team revisited the IARC report and found substantial differences between the draft version of the report on glyphosate's carcinogenicity and the final report that was filed. Specifically, the IARC had removed from the draft version important sentences that indicated "multiple scientists' conclusions that their studies had found no link between cancer and glyphosate in laboratory animals," and deleting from the conclusion a

115

sentence saying that the report "'firmly' and 'unanimously' agreed that the 'compound'—glyphosate—had not caused abnormal growths in the mice being studied" (Kelland 2017). In response, the IARC disputed all the claims made by the Reuters investigators, specifically noting that the purportedly deleted passages from their draft report were only excerpts from a Monsanto report that they were disputing (Wild 2018).

The EPA read through the deliberations and ultimately ruled, based on the reports they gathered, that glyphosate is *not* likely to be carcinogenic to humans. Thus, despite the IARC reports and despite the court cases that were settled on the basis of the IARC conclusions, the EPA sided with Monsanto's claim that glyphosate was safe for use at levels currently allowed in the United States.

Charles Benbrook, a physician activist who had been following the glyphosate debate carefully and had written previously on his perception that GE foods had increased the risk of glyphosate exposure to humans, reviewed the opposing conclusions. In an article published in *Environmental Science* titled "How Did the US EPA and IARC Reach Diametrically Opposed Conclusions on the Genotoxicity of Glyphosate-Based Herbicides?" he offered this observation:

> EPA and IARC reached diametrically opposed conclusions on glyphosate genotoxicity for three primary reasons: (1) in the core tables compiled by EPA and IARC, the EPA relied mostly on registrant-commissioned, unpublished regulatory studies, 99% of which were negative, while IARC relied mostly on peer-reviewed studies of which 70% were positive (83 of 118); (2) EPA's evaluation was largely based on data from studies on technical glyphosate, whereas IARC's review placed heavy weight on the results of formulated GBH [glyphosate-based herbicides] and AMPA [a breakdown product of glyphosate] assays; (3) EPA's evaluation was focused on typical, general population dietary exposures assuming legal, food-crop uses, and did not take into account nor address generally higher occupational exposures and risks. IARC's assessment encompassed data from typical dietary, occupational, and elevated exposure scenarios. (Benbrook 2019)

In other words, the opposing deliberative bodies were studying different things and relying on different kinds of evidence, so both could be correct despite coming to opposite conclusions. In the terms of my analysis, they were situated in different places in the swirl. In the midst of all of this deliberation, Monsanto issued a statement about the IARC's classification of glyphosate. "We want to be clear. . . . All labeled uses of glyphosate are safe for human health and supported by one of the most extensive worldwide human health databases ever compiled on an agricultural product. In fact, every glyphosate-based herbicide on the market meets the rigorous standards set by regulatory and health authorities to protect human health" (GMO Answers 2015).

To be sure, the debate over glyphosate has not simply been about whose science is correct or flawed but rather about the fact that we live in times when consensus has given way to the swirl under duress from chemical alter-life (Murphy 2017b). This alterlife is, again, as much a product of academic capitalism, agrochemical industrialism, and the need for consensus forms of science as it is a product of the fact that glyphosate has multiple potencies and different ways of making itself present in different places. Consensuses were changing direction, splintering and forming new subgroups that offered competing certainties, one that spun this way and another that way—some that argued for glyphosate's absolute toxicity at any level, another for its toxicity at levels deemed safe by the EPA, another for its safety at EPA-approved levels, another for its safety at any level, and still others (like Benbrook, above) that its toxicity could be seen (or not) depending on what kinds of objects one scrutinized in research laboratories and how they measured the presence or absence of glyphosate.

Multiple constituencies were forming and reforming around glyphosate. To this day, despite the fact that the IARC report is contested and that more and more funding has poured in to sponsor new research that shows glyphosate is *not* carcinogenic, the idea that it probably *is* carcinogenic has nevertheless been sustained as a key formation in the life of glyphosate among many of its constituencies. Most significantly, its potency as a carcinogen would give activists purchase in policy and legislative worlds, making this wayward and ubiquitous chemical a key interlocutor in their legislative activist efforts.

Glyphosate Navigates the Swirl

In June of 2017, building on the IARC report, activists from California attended a hearing at the California office of the US EPA called the Office of Environmental Health Hazard Assessment (OEHHA), one of the more progressive of all state-run health and environmental agencies. The topic under consideration was the level of no significant risk (NSRL) that had been proposed for glyphosate. After the OEHHA got the news of the IARC and WHO classifications, concerned legislators who had for many years been petitioned by anti-GMO food activists proposed to add glyphosate to its list of regulated chemicals under Proposition 65 (the Safe Drinking Water and Toxic Enforcement Act of 1986, also known as the "Right to Know" Act). According to a press release issued by the OEHHA on March 28, 2017, Proposition 65 requires the state to maintain a list of chemicals known to cause cancer, birth defects, or other reproductive harm. In its words, this proposition does not ban or restrict the use of listed chemicals, but, rather, requires businesses to provide warnings prior to causing a significant exposure to a listed chemical and prohibits discharges of the chemical into sources of drinking water.

The specific data that the panel referred to in its determination were taken from studies used by the IARC that produced evidence of glyphosate's potential carcinogenicity, based on these findings: "IARC's classification of glyphosate in Group 2A humans"), with a finding of sufficient evidence of carcinogenicity in experimental animals... with limited evidence in humans for the carcinogenicity of glyphosate . . . and a positive association for non-Hodgkin lymphoma" (OEHHA 2017a). This research, despite being contested by the EFSA, led the OEHHA panel to decide that there was sufficient evidence to warrant deliberation about glyphosate being included in the Proposition 65 list in a public hearing.

The proposed NSRL that was to be implemented through Proposition 65 was 1100 micrograms of glyphosate per person per day, or 1.1 milligrams based on a 70 kg person (Delson 2017). This level was determined on the basis of the studies used in the IARC report that showed increased likelihood of developing tumors (at rates higher than expected in study animals) at doses of 1000 micrograms or more as a dose response assessment (OEHHA 2017b). This NSRL would be 127 times lower than current US allowable reference levels but higher than European allowable levels. In other words, this chemical would go from being considered no risk (because unsafe levels were assumed to be high enough that most users would not be exposed to that amount from normal use of a product) to having a risk level associated with it that would have to be made public to anyone who bought a product containing glyphosate. Even before the hearings about this specific legislation, Monsanto filed a lawsuit against the state of California for proposing the inclusion of glyphosate in this list, apprehensive about any legislation that might come out of it and, critics said, hoping to tie up the implementation of any new regulations in court battles.

Soon after the proposal was made to OEHHA to add glyphosate to the Proposition 65 list, public hearings were held.[1] The mandate of the OEHHA hearing was to decide if current NSRLs were adequate, but most of the activists at the hearing were focused on a different question altogether: should any level of glyphosate be allowed? The most hopeful activists knew banning glyphosate was not only a way to get rid of Roundup; it was a first step toward getting rid of Roundup Ready foods and thus the first step to eliminating genetic engineering from the California food system altogether. This outcome was highly unlikely and would not mean that GE foods designed to be used with herbicides other than Roundup would not simply replace Roundup Ready crops, but this did not prevent activists from trying. Rounding up the troops, they prepared strategies for their full day at the hearing (one of several that would be held). Monsanto representatives at the hearing had the opposite goal: to prevent glyphosate from being listed among the chemicals in Proposition 65 and, if that was not possible, to make sure the NSRLs proposed were not lowered. They, too, had prepared testimony and supporting documents.

The hearing in June of 2017 was long and full. Among those who offered testimony was Zen Honeycutt, a mother of three children and founder of a nonprofit activist group committed to eliminating toxic pesticides called *Moms across America*. She had some of her own food tested and calculated the typical amounts of glyphosate eaten by a child each day in oatmeal, hummus and pita, milk, corn chips, berries, water, eggs, orange juice, toast and jam, and pasta.[2] Honeycutt summed up by reminding the panel that the total amounts of glyphosate in one day's consumption of food for the average (not food insecure) child was 2484.4 micrograms per person per day. In other words, according to her calculations, the average child in California was already eating more than twice the amount of glyphosate being proposed as the new NSRL.

A retired biochemist followed her. He reported on not only glyphosate's tumorigenic properties but also its capacities as a neurotoxin—an endocrine disruptor—and thus its links with infertility, thyroid disease, and brain damage. Another presenter spoke on behalf of farmworkers whose exposures to toxic pesticides were a social justice issue. Farmworkers, who were mostly Latinx, suffered higher rates of exposure to all pesticides and thus had higher rates of health issues from these exposures. Another presenter with a science background offered a deep dive into the biochemistry of glyphosate. It is a synthetic amino acid, an analog of our canonical amino acid glycine, a protease inhibitor ... he said, running his fingers through his gray hair, frustrated by the time constraints of only five minutes. You just don't mess with glycine, he warned, apparently unaware of the scientific publications that had shown glyphosate does not substitute for glycine in actively dividing mammalian cells (argued in Mesnage and Antoniou 2017). He reminded the panel that as a chelator, the suspicion was that it could be chelating important minerals in the body that are needed for health, such as iron, manganese, and zinc.

My colleague Michelle was there too, offering testimony of her own. Using the well-rehearsed narratives I had heard before, she described the kids she was seeing in her clinic as sicker than any generation prior. Kids who had been born into the generation of GMO foods were suffering from high rates of dysbiosis and leaky gut. She talked about new studies of glyphosate's association with nonalcoholic fatty liver disease in laboratory rats as well as glyphosate's antimicrobial potency and thus its probable disruption of the gut microbiome. She argued that changes at the epigenetic level were being passed onto children who are being born "pre-polluted," without the ability to get rid of toxicants in their bodies. She reminded the panel that we cannot track cancer in children; that we won't know until these kids are ten years older what these chemicals are doing to them. There are no safe levels of glyphosate, she said, reminding the audience that she could test for it, treat it, clear it, and kids get better. . . . Using a familiar line, she said that we are all facing nothing short of a public health disaster.

The Monsanto representatives, some of whom were lawyers, were also prepared. They listed a variety of consensus scientific studies that showed no carcinogenic effects from glyphosate-based herbicides and asserted that the current NSRLs were sufficient to protect the public from any putative risks. They also reminded the panel of the risks to the agricultural sector of removing this vital and safe herbicide. They were followed by a young woman, a farmer from California's Central Valley, whose family had been farming for four generations. She appealed to the panel's sense of reason by asking them why she would continue to use glyphosate-based herbicides on her farm if she thought it would harm her children or family. Obviously, she would not.

Factual claims about glyphosate were flying about with certainty from credentialed and noncredentialed activists, performing emotional and sensible as well as scientifically reasoned appeals about the risks and safety of this chemical. The presentations ended with two important testimonials. The first was from a middle-aged man who walked slowly to the podium, giving the impression he was in fragile health. With a shaky voice he told the audience that in his twenties he was an avid user of Roundup. He had it around his home and used it all the time. He always took precautions—if he got it on his skin he washed it off right away. He "followed the precautions," he repeated, but remembered sometimes getting nausea after using it. In 2010 he was diagnosed with non-Hodgkin's lymphoma. He told the audi-

120

ence he didn't know how he contracted it, but since that time (and his voice began to crack) the quality of his life had totally degraded. He wondered how ordinary people like him could be in a situation in which they were being exposed to a toxic chemical without knowing it. He finished by reminding people that he was there to protest the use of glyphosate. I felt for him and although he said he didn't really know how he got the cancer, I too suspected that it must have been the Roundup.

He was followed by an elderly woman who told us about her grandchild, who was born with neurological defects (low tone cerebral palsy). Her granddaughter cannot sit up; she cannot swallow. She can't chew or ingest without aspirating her food. She became very emaciated and was put on a G-tube. The grandmother said that she is still being fed that way "because her mom and dad do not understand about glyphosates." Her grandchild was prescribed PediaSure. "It comes in a can. It is very high in glyphosate." She couldn't remember if it was "fifteen times or fifteen thousand times the allowable level in Germany." She held her composure as she tried to tell the audience about the impact on her family and how it is a crime that so much glyphosate is in so much baby food. "People have no idea how dangerous it is," she said. "It should at least be labeled so people will know they are poisoning their children." I also wanted to tell this panel that they should ban glyphosate. The emotional depth of despair and sense of foreboding being presented to the panel nevertheless gave the activists a surfeit of hope. Their pleas formed an affective surge of sympathy and certainty that, at a minimum, OEHHA could not ignore.

Soon after the hearing, I accompanied some of the same presenters on a trip to lobby California legislators over a bill to prohibit any use of glyphosate in parks, schoolyards, and day care centers. California already had a bill prohibiting the spraying of pesticides near schools, playgrounds, and day care centers between the hours of dawn and dusk. But this policy had only exacerbated the problem of children being exposed, since groundskeepers made sure to spray in the early morning hours just before children or visitors arrived, making it more likely that people would be exposed at the most saturated times of day. The bill this group was mustering support for would ban the use of glyphosate at these sites entirely. One of our activist team had already succeeded in getting her entire city to ban the use of pesticides, including and especially Roundup, near these sites. We began the day on the road, jammed into a hatchback.

Our group was hosted by another nonprofit organization that was working to protect farmlands and eliminate toxic pesticides. Glyphosate was

121

their primary concern that day. They had lined up five meetings: four with California state senators, one with an assemblyman, and one with an aide to Governor Jerry Brown. As we drove up, we plotted our strategy. "We need to start with the fact that glyphosate is dangerous," one member of our group said. "We can also talk about less toxic alternatives to Roundup. We should mention that other places have banned glyphosate altogether." Our driver, a farmer from the Midwest who had flown out for the day's events (and whose own grandchild was born with severe health problems), said that there were organic vinegar-based alternatives to glyphosate that were just as effective but less toxic. "They [the legislators] are probably confused and are hearing so much from the other side on the cost-effectiveness for landscaping and crops," he said. "So we need to remind them that the bottom line [for banning it] is defensible with the data we have. We can tell them that there are cost-effective alternatives."

Another of our team wanted to use a different tactic. "We are going to be confronted with the typical roadblocks. 'Just Label It' [the campaign to enforce labeling of GE foods] pushed hard but the Farm Bureau reps had a lot of science that swayed the vote, even though they [the activists] were told that they [the legislators] were sympathetic." Shifting tactics, he then said, "We need to discredit their objections. Go into the Monsanto Papers and [find] all the evidence that glyphosate is carcinogenic, that it causes reproductive problems, that the company has been funding disinformation campaigns. They [Monsanto] were caught with their pants down." He was referring to the trove of internal documents that had been made available through trial discovery showing Monsanto's deliberate campaign to silence the factual information that ran counter to the industry consensus about glyphosate's safety and harm, among other things.[3]

Another of my fellow travelers added: "We need to use marketing strategies, use 'sizzle points' ... like ... that it is a carcinogen, causes birth defects, autoimmune diseases, mental health effects. We need to hammer home the health deficit from glyphosate." Everyone in the car had a different idea about the best tactics. Michelle added her opinion: "Chronic disease is rampant. People think it is normal to have these problems. But you can change them with dietary changes. We can offer them the health of our kids." Our ringleader and driver returned to our specific mission: "We need to hone in on the ask: trying to get it banned in cities and parks, schools, where children play. Our strategy should be, 'It's what people want. It's happening elsewhere [banning it] and it is possible and successful. There is already a movement afoot. It's just like lead and tobacco.'"

After we arrived at the capitol we began a madcap day of hurried visits, being shuttled from one office to the next, waiting for someone to show up, having to hold some meetings in the hallway because our visit had been forgotten. One member of the assembly actually sat down and listened to our concerns, but most just sent their assistants to dutifully hear us out. They listened carefully. They nodded in assent to the idea that we should be worried about protecting our children and should find alternatives to toxic sprays. Some of them had not yet heard that glyphosate had been found to be carcinogenic. Were we talking about Roundup? Was that the only concern?

Although no assembly person actually committed to supporting the bill outright that day, and even though the office of the governor took the time to schedule us in and listen to our cause, we all felt rather demoralized in our effort when the day ended. One of the aides to the Majority leader of the Senate told us that to get this sort of bill passed we would need a large constituency behind us, like the California Nurses Association or the United Auto Workers. The farm constituencies were very powerful in our state, she said. It was unlikely a bill like this would pass without the support of the farmers. The effort to ban the use of glyphosate products near schools, childcare centers, and playgrounds went nowhere.

Still, months later, after much deliberation, the California legislature honored the decision of the OEHHA hearing that recommended including glyphosate in Proposition 65 under the allowable levels of 1100 micrograms per person per day. The panel spent a good deal of time producing their final report and responded in some detail to all of the concerns expressed at the hearing, even the testimonial claims that were at best ill-formed. They drew on the archives of published journal articles and metastudies that enabled them to dismiss the majority of concerns raised at the public hearings except the one about glyphosate's probable carcinogenicity. By adding it to the list of chemicals covered by Proposition 65, the legislature determined that the makers of Roundup and any other glyphosate-based herbicide would, as of 2017, have to provide warning labels about its carcinogenicity on any products sold in California. Not one legislator ever made any suggestion that banning the chemical would be a good idea. By 2020 Monsanto had sued the California legislature and found a judge who would overturn the OEHHA decision, making it impossible for the state to enforce the labeling requirement (Arcuri and Hendlin 2019; Steptoe 2020). But in many ways, the momentum could not be stopped. Whatever rulings were happening in the legislative and regulatory spheres, the swirl had already launched glyphosate into the deliberative space of the California courts.

123

Like the NASEM, the IARC, the EFSA, and the various company reports and scientific metastudies that have been produced to speak glyphosate's truths, the final report from OEHHA also had to use tactics of evidentiary scale-up to bridge the gaps of knowledge about how carcinogenic, if at all, glyphosate actually is. But what happened in the OEHHA process was an orchestration of proliferative claims around one chemical and that chemical produced a world of narratives of its own, perhaps a scale-free enactment of a swirl in motion. Glyphosate took center stage, enabling its human interlocutors to speak about its potencies in their lives and in others' by calling out its multiple effects and multiple harms, culminating in one harm to rule them all: carcinogenesis. Glyphosate became the master narrator and an active participant in a political turmoil of truth claims that swirled this way and that and then settled for a moment on its potencies as a cellular mutagen. In many ways, the mere proliferation of debates about glyphosate—even the competing claims about its possible carcinogenicity—have made it a useful ally to activists, indeed one might say an activist in and of itself. This is even more clear in its travels to courtrooms where the legal recognition of its harm has been deliberated.

Glyphosate's Accountabilities

In August 2018, Dewayne Johnson, a forty-six-year-old former groundskeeper for a public school in Benicia, California, won his lawsuit against Monsanto on grounds that the company had failed to warn him of the dangers of glyphosate in causing his non-Hodgkin's lymphoma. Mr. Johnson testified that he had used this product liberally for years to eliminate weeds at the school without knowing it was potentially harmful. The product he used was ProRanger, Monsanto's generic brand of Roundup, which contained 41 percent glyphosate (the same amount as Roundup).

Mr. Johnson described his feeling of having been misled by the company, telling the court that "he wore protective gear while spraying to be extra cautious. But there were occasional leaks, and one time his skin accidentally became drenched," according to one reporter (Levin 2018). Another journalist reported that Mr. Johnson "called a Monsanto Co. hotline twice—once before his diagnosis, once after—and asked whether the herbicide he was spraying on the job, the most widely used weed killer in the world, could cause harm to humans. Both times, [he] said, the person at the other end of the line listened to his account of being accidentally doused with the herbicide glyphosate, and said someone would call him back. No one ever did" (Egelko 2018). Later, in another interview, Mr. Johnson gave more details:

That day of the accident, the day the sprayer broke and I got drenched in the juice, I didn't think that much about it. I washed up in the sink as best I could and changed my clothes. Later I went home and took a good long shower but I didn't think, "Oh my god, I'm going to die from this stuff." Then I got a little rash. Then it got worse and worse and worse. At one point I had lesions on my face, on my lips, all over my arms and legs.

When I first saw a doctor he was totally confused and didn't know what was happening on my skin. He sent me to see a dermatologist who did a biopsy of a lesion on my knee. They sent me to UCSF [University of California, San Francisco] and then to Stanford. A bunch of doctors came and checked me out. Then one day I got a call. They told me it was urgent, I had to come in to discuss my test results. When the doctor said I had cancer, my wife was sitting there with me. She started crying. I didn't take it in right away. I don't think I have still taken it in.

People want to say it's Johnson v. Monsanto. They want me to talk about the company. I don't want to do that. I don't even want to say the company name. I just say "the big company." I don't want to be slanderous. I've seen reports that I want an apology but that's not true. I'm not a person who would think an apology would make me feel better—it certainly would not heal my cancer. This isn't about me and that big company. It is important for people to know this stuff, to know about what they're being exposed to. If people have the information they can make choices, they can be informed and protect themselves. I'm just a regular guy from a small town called Vallejo in the California Bay Area who happened to seek the truth about my failing health and found answers.... I would never have sprayed the product around school grounds or around people if I thought it would cause them harm.... They deserve better." (Gilham 2018)

The potential effects of Dewayne Johnson's lawsuit were many, including reiterating the need for better product labeling. But in this case, the decision went further against Monsanto in awarding punitive damages. The company was ordered to pay Mr. Johnson $289 million. Reporters insisted that it was the fact that the judge allowed scientific reports from organizations like the IARC to be used in the case that convinced him of Monsanto's liability.

Of course, Monsanto immediately appealed the ruling, surely hoping to stem the ocean of other plaintiffs across the country who were lining up to sue the company for similar wrongdoing. Monsanto lawyers brought in

125

counterevidence, the mountains of evidence and reports from the NASEM and from scientists who contested the findings of the IARC to show that glyphosate was, in fact, not the culprit for Mr. Johnson's cancer. Two months later, a judge reduced Mr. Johnson's award to $78 million, not on grounds that Monsanto was not liable but simply on grounds that the ratio between compensatory and punitive damages must be one to one in cases like this. Mr. Johnson accepted that settlement, noting that in all likelihood he would be dead before any payout was seen (Sullivan 2018).

Not quite a year after Dewayne Johnson's case, a second one was brought against Monsanto by another California man who had non-Hodgkin's lymphoma and who blamed his disease on his extensive use of Roundup to manage the weeds in his fifty-six-acre ranch. Edwin Hardeman, then seventy, "had proven the herbicide was probably a 'substantial factor' in causing the cancer with which he was diagnosed in 2015. Mr. Hardeman said he'd sprayed Roundup on his property for decades" (Egelko 2019). Although Mr. Hardeman's cancer was in remission, unlike Dewayne Johnson's, he "testified that he sprayed Roundup for nearly thirty years to kill poison oak, often feeling the liquid on his hands or inhaling it" (Egelko 2019).

The judge in Edwin Hardeman's case noted that the evidence linking Roundup to cancer was "shaky" but rejected Monsanto's motion for dismissal and ruled that the evidence was sufficient to bring the case to trial. He split the trial into two phases, the first to consider whether the evidence was sufficient to ascertain that Roundup was a likely cause of Mr. Hardeman's cancer, and the second to consider Monsanto's liability for damages. Using the massive supply of materials that had surfaced in the year prior to the case that provided evidence that Monsanto concealed information about glyphosate's potential toxicity (that is, not evidence from new science on the toxic effects of glyphosate in laboratory animals but evidence that Monsanto hid information from regulatory agencies or otherwise obfuscated information about the level of toxicity of the chemical), Edwin Hardeman's lawyers argued that "it is clear from Monsanto's actions that it does not particularly care whether its product is in fact giving people cancer, focusing instead on manipulating public opinion and undermining anyone who raises genuine and legitimate concerns about the issue" (Egelko 2019). On the first issue—whether it was likely that his cancer was caused by Roundup—Mr. Hardeman was awarded $5 million. On the second issue, decided in July of 2019, the initial request for $75 million (Monsanto's liability for damages) was granted, but reduced to $20 million. After these cases, more lawsuits against Monsanto on behalf of people with non-Hodgkin's lymphoma began to pop up.

126

Monsanto continues to appeal the rulings against it, but this has not prevented more lawsuits from moving forward on behalf of people with non-Hodgkin's lymphoma with similar stories about their long-term use of Roundup or related glyphosate-based products. The American Society of Clinical Oncology predicts that 74,200 Americans will be diagnosed with non-Hodgkin's lymphoma by the end of 2020. At last check, there were five thousand to seven thousand lawsuits against Monsanto that were lined up to be pursued on behalf of people with non-Hodgkin's lymphoma. An advocacy group called Roundup Cancer Claim has set up a website for inquiries about pursuing legal action against Monsanto for people who have cancers including non-Hodgkin's lymphoma, large B-cell lymphoma, follicular lymphoma, hairy cell leukemia, mantle cell lymphoma, Burkitt lymphoma, peripheral T-cell lymphoma, skin lymphoma, and others (US Right to Know n.d.b).

However promising these cases feel, it would be naïve to overestimate the prospects for court-deliberated rewards to the public in cases like those of Dewayne Johnson, Edwin Hardeman, or the many others who are lined up with lawsuits. The outcomes will likely result in no more than settlements to victims or their families and slap-on-the-wrist fines for the company, if settlements are actualized at all. The fact that these legal cases were concluded in favor of the plaintiffs who were able to convince juries and judges that glyphosate had caused non-Hodgkin's lymphoma does not mean that the truth about glyphosate has settled. Glyphosate offers a good example of how a chemical can pull the swirl (perhaps temporarily) into alternative formations, fracture the consensus, and fugitively find new grounds for habitation in bodies, soils, and foods, and redress for dying people and their families. Glyphosate's potencies continue to grow, even as the deliberation over its potencies continues to foil uniform policy changes.

Glyphosate Outlives Its Human Companions

In 2017, a trove of Monsanto documents about glyphosate was made available and space was found for them to be housed at my university (with support from Stanton Glantz, the professor who acquired the famous tobacco archives). One of the documents was a letter sent in 2013 by Marion Copley, who worked at the EPA, to Jess Rowlands, the head of the EPA at the time. Her letter has now circulated widely, even after her death from cancer in 2014 at age sixty-six, partly because it is thought to reveal the hidden conspiracy that many activists believed was taking place between Monsanto and US regulatory agencies. It is singularly and unapologetically tragic.

127

Jess,

Since I left the Agency with cancer, I have studied the tumor process extensively and I have some mechanism comments which may be very valuable to CARC based on my decades of pathology experience. I'll pick one chemical to demonstrate my points.

Glyphosate was originally designed as a chelating agent and I strongly believe that is the identical process involved in its tumor formation, which is highly supported by the literature.

- Chelators inhibit apoptosis, the process by which our bodies kill tumor cells
- Chelators are endocrine disruptors, involved in tumorigenesis
- Glyphosate induces lymphocyte proliferation
- Glyphosate induces free radical formation
- Chelators inhibit free radical scavenging enzymes requiring Zn, Mn or Cu for activity (i.e. SODS)
- Chelators bind zinc, necessary for immune system function
- Glyphosate is genotoxic, a key cancer mechanism
- Chelators inhibit DNA repair enzymes requiring metal cofactors
- Chelators bind Ca, Zn, Mg, etc to make foods deficient for these essential nutrients
- Chelators bind calcium necessary for calcineurin-mediated immune response
- Chelators often damage the kidneys or pancreas, as glyphosate does, a mechanism to tumor formation
- Kidney/pancreas damage can lead to clinical chemistry changes to favor tumor growth
- Glyphosate kills bacteria in the gut and the gastrointestinal system is 80% of the immune system
- Chelators suppress the immune system making the body susceptible to tumors

Previously, CARC concluded that glyphosate was a "possible human carcinogen." The kidney pathology in the animal studies would lead to tumors with other mechanisms listed above. Any one of these mechanisms alone listed can cause tumors, but glyphosate causes all of them simultaneously. It is essentially certain that glyphosate causes cancer. With all of the evidence listed above, the CARC category should be changed to "probable human carcinogen." Blood cells are most exposed to chelators, if any study shows proliferation of lymphocytes, then that is confirmatory that glyphosate is a carcinogen.

128

Jess, you and I have argued many times on CARC. You often argued about topics outside of your knowledge, which is unethical. Your trivial MS degree from 1971 Nebraska is far outdated, thus CARC science is 10 years behind the literature in mechanisms. For once in your life, listen to me and don't play your political conniving games with the science to favor the registrants. For once do the right thing and don't make decisions based on how it affects your bonus. You and Anna Lowit intimidated staff on CARC and changed MI ARC and IIASPOC final reports to favor industry. Chelators clearly disrupt calcium signaling, a key signaling pathway in all cells, and mediates tumor progression. Greg Ackerman is supposed to be our expert on mechanisms, but he never mentioned any of these concepts at CARC and when I tried to discuss it with him he put me off. Is Greg playing your political games as well, incompetent or does he have some conflict of interest of some kind? Your Nebraska colleague took industry funding, he clearly has a conflict of interest. Just promise me not to ever let Anna on the CARC committee, her decisions don't make rational sense. If anyone in OPP is taking bribes, it is her.

I have cancer and I don't want these serious issues in HED to go unaddressed before I go to my grave. I have done my duty.

Marion Copley March 4, 2013 (M. Adams 2017)

It might be tempting to read these events as a typical case of corporate plunder and regulatory capture, the culmination of many years of unethical behavior on the part of Monsanto and the regulators it "bought." We could argue (as others have; see Robin 2010) that companies like Monsanto anticipated the potential for harm from their chemicals and have built in payoffs like this to their bottom lines and measured the costs to settle lawsuits against the huge profits made by these products. This may also be true. But I think glyphosate can help us think through and beyond this sort of critical engagement—an engagement that, to be realistic, has done little to help us get rid of chemicals that we know may be causing harm. Glyphosate trips up even the most heart-wrenching commitments to hold chemical companies accountable by its capacity to be everything and nothing at the same time in the games of certainty about harm. The list of concerns raised by Marion Copley had, within a few short years of her letter, already become points of contention and dispute in the scientific and activist communities that cared about this chemical. Thus despite the deaths of some, and perhaps many others, from cancers that may or may not have been caused by glyphosate, we are left with a world full of glyphosate, swirling and swooping and seeping into soils and water and plants and foods and cells, where it continues to swirl in formations that, scale-free, sometimes cluster into what looks strikingly like evidence of harm.

129

8

Chemicals as Agents of Care

Glyphosate has been enlisted as a partner to humans in various pursuits. Chemical scientists have benefited from glyphosate's capacities as a weed killer and partner to genetically engineered foods. Scientists have conscripted glyphosate to counter the scientific consensus produced by agrochemical industries. Activists have enrolled it as an ally in their pursuit of government restrictions on its use. Humans have crafted specific stories about this chemical in ways that construct plausible narratives about injury and responsibility. What activists do with glyphosate is possible in part because of what glyphosate does for them. Its abilities to seep into different tissues and infrastructures enable it to alter bodies and environments. Glyphosate is all these things—present in all these places; altering the things it touches; pushing conversation and politics to one focal point, then the next; and leaving behind new laws, even if also ongoing possibilities for harm. In this sense it operates as an agent of some sort of care, partnering with different physical, emotional, and political constituencies in their efforts to change things.

From the perspective of this chemical, however, we could argue that humans are not end points in glyphosate's life, but rather stopping points as it marches on through environments and bodies, swirling this way and that through tissues, soils, scientific laboratories, and courts of law. In one sense, glyphosate has the capacity to be transmuted from one material life to another as it moves from sprays to soils to plants to foods to bodies to cells to genes. This transmutability enables us to see its swirl-like movements, much the way Annemarie Mol (2008) describes the agency of an apple she eats. Nutrients from the world outside that have been absorbed from the soil and stored in the apple's flesh are delivered with each bite into her body, where they break down into various metabolites that keep the blood pumping, the skin smooth, the digestive system adequately discerning what is waste and what is worth keeping. Is the agency in the I or in the apple? she asks. She goes on: isn't the apple also the farms and fertilizers and seeds and shipping companies and grocers enabling it to be purchased for consumption by me? Couldn't we consider all of these transubstantiations part of the agency of the apple? Glyphosate is involved in similar kinds of agentive transubstantiations—producing situated and material effects that are multiple. The apple is both one thing and many things, like glyphosate.

Hannah Landecker (2019) offers another version of the multiplicity of chemical agents in her tracing of arsenic, a naturally occurring toxic metal that appeared in the US food ecosystem starting in the 1950s. Finding its way from slag heaps as a waste chemical into livestock food supplies, where it gave food animals a rosy and fatty appearance, arsenic is an active metabolic agent in an industrial ecosystem in which inputs and outputs are managed to maximize commodities and eliminate waste. This industrialization in the case of arsenic was about not just building ships and roads and material products but also the reshaping of food systems as sites for the productive reuse of waste products—a technology that Monsanto played a significant role in developing. Given the optimism about chemicals like arsenic, and now glyphosate, it makes sense that chemical resources would find their way from one industry to the next; what is waste in one system has a use in another. Arsenic residues showing up in the bodies of those who ate these foods could thus be seen as part of the metabolic life of this chemical. First a chelator, then a weed killer, and now an integral ingredient in the industrial crops that become food, glyphosate has had an interesting journey. Perhaps non-Hodgkin's lymphoma and malfunctioning guts and immune systems are one irruption of glyphosate-rich foods in bodies—bodies that are conduits for chemical life. Glyphosate absorbed through guts can travel to lymph, where it is probably associated with higher than usual rates of creation of micronuclei— fragments of chromosomes that are not incorporated into a daughter nuclei during cell division, genotoxic events augmenting chromosomal instability, otherwise known as cancer in the form of non-Hodgkin's lymphoma.

For Mol and Landecker, the agentive thing (apple or arsenic) helps me understand how things are situated in their material effects. My sense of

131

glyphosate as a chemical with agency works along the same lines, but I am also arguing for it as an agent of care in other ways, too. Glyphosate is an element of a swirl that enables it, ironically, to care in a political sense—as a technology of politics—for humans. By this I mean something along the lines of what Murphy (2015) refers to as an *unsettling* form of care that, in their work, is aimed at disrupting sedimented arrangements of historical, racialized, and gendered valuations and devaluations that are so often reproduced in efforts to help. Might glyphosate disrupt the notion that care must not be violent? If we are to believe what its particular potencies in this sector of the swirl would have us believe about its carcinogenesis and its injury at multiple levels of the living landscape, then glyphosate harms even while being pulled into a swirl of efforts to care.

Glyphosate shakes things up and reorganizes sensibilities for life, experience, and harm. This might be as true for bodies that now live with glyphosate as it is for clinical efforts to sleuth its role in disrupting biological systems or in rethinking environmental survival under the pressures of agrochemical industrialism. María Puig de la Bellacasa points to this form of care when, in her reading of the politics and engagements of regenerative agriculture, she notes that humans are not and should not be the centerpiece of our thinking about how to reenvision a future for planetary survival in relation to agricultural ecosystems. Just as chemicals like glyphosate can stimulate thinking about what it means to engage critically in imagining a future without their ubiquity, she thinks about how to reverse climate change through animations of the soil. In what she calls the *foodweb*, humans are "full participant 'members' of the soil community rather than merely . . . consumers of its produce or beneficiaries of its services" (Puig de la Bellacasa 2017, 192). Permaculture (a specific kind of regenerative agriculture) is an antidote to agrochemical industrial capitalism, and in this farming system the soil is not something we take care of, but rather something that we take part in. Soil takes care of us as it displaces humanity from the center of a sustainable ecosystem. In her view, participation in the foodweb is an invitation to live in a world in which human survival is not the beginning or end of the story. This care points to an ethics that "cannot be about a realm of normative moral obligations but rather about thick, impure, involvement in a world where the question of how to care needs to be posed"—and, I would add, posed over and over again (Puig de la Bellacasa 2017, 6).[1]

Glyphosate is an ambivalent agent of care. It has revealed how agrochemical capitalism depends on a certain expenditure of life even while claiming to sustain life via toxic pesticides, but it has also become a pivotal chemical in the swirl of efforts to change these things, being an anchor and provocateur for legal adjudication over accountability for chemical harm. The legacy of glyphosate in this arc and the engagements it has had along the way have produced dead bodies of microbes and insect pests and weeds that are scattered about the depleted soils of agroindustrial empires, along with (perhaps prompting) an epidemic of chronic ailments, dysbiotic guts, and immune systems crowding the hallways of American health clinics. In addition, the credibility of the scientific consensus is also now strewn about the archives of industry-funded and academic research like the detritus of a scientific empire that has long since deserved to collapse. Like the dust that gets lifted up by the swirl that seems to hold purchase, then doesn't, then directs us to other sites of action and effect, glyphosate has been an agile interlocutor for thinking otherwise.

What can be said, then, about the ethical opportunism of the swirl? What acuities of glyphosate's movements are brought into relief at some points and diffracted by the pull of inertia and harm that becomes visible elsewhere? I am convinced that the swirl effectively captures our current predicament with chemical harm, with late industrial scientific research, with agrochemical industrialism, and with the activism around environmental injury. Like the movements of the chemical glyphosate, the swirl clusters and swoops, settling here and then there, operationalizing new political opportunities as it goes. It refuses a politics that insists on the facts being immutable, just as it refuses a politics that suggests the facts can ever speak for themselves. The swirl does not refuse the potencies of the chemical or the actions that the chemical has enabled, but offers a way to trace these that allows us to recognize how to place commitments in these unsettled assemblages. My sense is that this is not just an analytical but also an ethical opportunity. To see this, I return one last time to the fortunes of glyphosate in its recent corporate life, which means returning to Monsanto, not because this company deserves special flogging but rather because the flourishing and diminishment of glyphosate in this swirl can help me orient my commitment.

Moving through the Swirl

In 2013, Monsanto was named one of the one hundred best corporate citizens by Corporate Responsibility Magazine. The same year, it announced plans for a $400 million expansion of Chesterfield Valley Research Center, including thirty-six new greenhouses, additional offices, and laboratory space as well as additional plant growth chambers to facilitate development of its seed and trait pipeline. It planned to add 675 jobs over the subsequent three years.

The vision was mapped by Hugh Grant, the head of Monsanto in 2006, who explained this forward-looking aspiration in an interview with Ray A. Goldberg from Harvard: "Our challenge goes back to our origin as a chemical company that sold a gallon jug of a product that kills weeds. Over the past several years, we've found ourselves in the corner offices of major food companies talking about future opportunities. They aren't simply CEO to CEO. They're technology-to-technology and fieldsman-to-fieldsman discussions. It's two different industries. Our cycle times are seven years; their cycle times are often scanner data over bar code data over one weekend. Our strength is on issues that need real technological solutions" (Goldberg 2018, III).

For Grant, the issue is how to create value: "Creating value beyond the underlying commodity is the oil that lubricates the system" (Goldberg 2018, III). He didn't talk about it in 2006, because it wasn't yet in the pipeline, but by 2014, Monsanto had joined other companies in upping the stakes of creating value by continuing to modify the genes of plants and animals in ever more creative ways. Juan Enriquez, president and CEO of Biotechonomy (also interviewed by Goldberg), offered a stunning example of the rationale and future imagined in these sciences that takes us back to the early days of Schneiderman's research at Monsanto and the company's interest in carving an ethical path for its chemical products:

There's an enormous overlap in the gene instructions between ourselves and a whole series of different species. The difference between ourselves and mice is about 5 percent, between ourselves and monkeys is about 1.27 percent. There are significant overlaps in gene code between ourselves and plants and even bacteria. The instruction sets that allow us to produce, synthesize and modify organics in plants may also be applicable to bacteria, and vice versa. We're starting to see plants producing vaccines and bacteria growing organic plastics, and vice versa. A cholera vaccine can now be programmed into tobacco, or a potato, or a banana or a bacterium.... [All this, he says] will push down prices of commodity products.... The biotech revolution is allowing farmers to code life or move from farming to pharming, or store information in living things or produce plastics. A few already produce broccoli that helps fight cancer, but that's the exception. Right now we have a system where "all-natural" corn might sell for $2, and "genetically modified corn" only $1.50,

134

because consumers tend toward all-natural. But someday the sign won't read "genetically modified corn"; it'll say corn that helps you fight cancer or process certain fats and the modified corn will sell for $4. Eventually the value added to plants and animals will have more consumer pull. (Goldberg 2018, 131–32)

In what might be considered a stellar example of the swirl, Grant's vision was to harness the power of genes-as-chemicals to rescue the population from rising health morbidities that chemicals like glyphosate might be causing. Turning to plants as medicines reverses the traditional ouroboros of agrochemical pharmaceutical empires. To bring this vision into reality, in 2015, the German company Bayer AG, mostly known as a maker of biotech pharmaceuticals, initiated the purchase of Monsanto. Bayer was the first chemical company to develop organophosphate pesticides in 1937 as part of its Nazi war effort (Walker 2010, 59). By 2018, the purchase of Monsanto was complete, with Bayer indicating it would retire the Monsanto name. It would, however, continue to produce Monsanto's agricultural products, including its GE seeds, pesticides, and fertilizers, and invest in a new stream of products in which plants could be genetically modified to become "medicines." This is in addition to reconceptualizing the agricultural enterprise in ways that prioritized the microbial health of soils by developing agrochemical additives in the same way they designed pesticides. With the merger of Monsanto and Bayer, the potential for double-dipping on profits was harnessed: the company whose chemicals flow through our food into bodies that, in one iteration of the swirl, are causing an epidemic of chronic disorders and cancers, in another turn would then also provide an arsenal of pharmaceutical drugs to treat them.

Still, the effects of the swirl are powerful. Soon after the first lawsuits against Monsanto were launched by people with non-Hodgkin's lymphoma in the United States, and just after Bayer bought Monsanto, Bayer's stock dropped 45 percent. By 2021, "Bayer shareholders in Germany would sue the company, claiming they should have been warned of the risk of lawsuits when the company acquired Roundup with its $63 billion purchase of St. Louis–based Monsanto Co." (Edwards 2021). Glyphosate, here, is a stock market provocateur. Bayer AG maintained that it did proper due diligence before the purchase and should not be held liable for the stock performance in relation to Monsanto. Deliberation over whether glyphosate should be blamed for the many cases of non-Hodgkin's lymphoma continue as this book goes to press. Juries and judges all over the US continue to have to decide which science to believe, whether the company must be held accountable for injuries it might have been able to anticipate (or could it?), if these diseased individuals had followed the safety guidelines on the pesticides' labels that warned against drenching oneself in the liquid (or had they not?), if this cancer could not have been caused by any other kind of carcinogenic chemical exposure (or did that even matter?), or whether the micrographic images showing carcinogenic cellular processes are being read accurately by scientists (or not?).

In early February 2021, before settling the lawsuit with Monsanto, Bayer decided to set aside first $2 billion then later $16 billion to pay for future litigation regarding claims of glyphosate being responsible for causing non-Hodgkin's lymphoma (Hals and Bellon 2021). Because of all this, Bayer has said, as of the writing of this book, it will stop selling Roundup and other glyphosate-based products to US consumers by 2023, but it will continue to sell them to commercial applicators and farmers indefinitely. It has also said it will gradually replace the combination of Roundup and Roundup Ready seed products with other pesticide-ready GE foods and pesticides. Considering Bayer spent $63 billion to purchase Monsanto and its gross income in 2014 was over $40 billion according to its investor web page, the costs of settling whatever lawsuits against Monsanto are successful in the years to come and of refining its market priorities for the product will do little to eliminate glyphosate (or more toxic pesticides) from the US agricultural market. All of this could change before this book goes to press. Glyphosate moves on, swooping and clustering and settling, then moving to new sites, even as it is displaced by other chemicals. In a world without glyphosate, we would still be pushed and pulled, drawn into and extracting ourselves from the swirls it brought to life. Roundup is still marketed to Americans.

Given all this, you may be wondering why I have chosen to return here to a somewhat cynical critique of the machineries of agrochemical and pharmaceutical capitalism and the hope that these companies will hold themselves accountable for any chemical injury they have caused. You might also be worried that I am suggesting that we should hold out for a massive shift to permaculture and regenerative farming to solve the problem of chemical harm from food, as if these approaches will be able to escape the swirl of contested science under the duress of agrochemical empires that has already formed in relation to them.

To be sure, the idea of agrochemical company accountability, just like the idea of large-scale shifts to regenerative agriculture, is a desirable one to hold onto. But that is not my goal. I have tried to follow glyphosate into the many places it has become visible, the material and conceptual presences it has, letting its multiplicity and its ability to spread as if in the atmosphere in different ways to different places. I have tried to let glyphosate become my guide.

In this journey, I've shown how glyphosate has played an important role in altering the structure of academic capitalism in relation to our food systems. It has played an important role in reorganizing crops, weeds, and chemical dependencies for farmers. It has also helped spark a flurry of alternative thinking in clinical care—a sensorium of possibilities for explaining heterogenous chronic morbidities that all relate to the diverse ways glyphosate penetrates and disrupts biological systems. Glyphosate has wreaked a kind of havoc with the utopian dream of a scientific consensus. I have argued that it offers a particularly clear lens on the predicament we have gotten into in relation to the scientific archives on chemical harm and that efforts to trace things like certainty and consensus instead lead to endless formations of swirl. I have tried to explore how that swirl—a conceptual architecture for thinking about life with chemicals—can augment our approach to the science and experience of chemical injury. And I have argued that we might think of glyphosate, in its grand finale, as taking up the role of activist as it is recruited to the work of shifting the consensus in efforts to legislate new regulations about its use. If you have developed a vertigo about glyphosate and its presence in the world, then I have done my job.

Glyphosate invites us to get comfortable with the movements of the swirl with a distributive politics that pulls us toward greater or lesser immersions in glyphosate's many worlds, from the cellular all the way to the grand social and infrastructural arrangements of capitalism, research, regulation, clinical care, and so on, pulling us along with it toward one end point and then another in ways that augment and empower this little chemical to be much more than it ever could of its own accord and, in so doing, sometimes thwart us, again and again.

Chapter Two. Building the Food Chemosphere

1. For a visual tour of the ride, see Monsanto, "Yesterland: Adventure thru Inner Space, Presented by Monsanto" (web page), updated November 1, 2019. https://www.yesterland.com/innersp.html.

2. I am grateful to the well-known research biologist (and interlocutor for anthropologists) Scott Gilbert for sharing his insights about Howard Schneiderman. Gilbert was a Howard A. Schneiderman Professor Emeritus at Swarthmore College.

3. Genetically modified mice were created by Rudolf Jaenisch in 1974. In 1973 research in recombinant DNA by Herbert Boyer and Stanley Cohen built on the techniques developed by Paul Berg for inserting viral DNA into bacterial DNA, creating new strains of bacteria (Rangel 2015). Once again, the discoveries quickly led to deliberations over the ethics of this technology and, despite a moratorium on GE experimentation that lasted for roughly one year (resolved with ethical guidelines at the famous Asilomar Conference in 1975), the US government began allowing the patenting of living organisms. The commercialization of the technology soon led to the development of genetically modified bacteria that could produce somatostatin (used in rBST) and then insulin in 1978. But this was not the same as producing living organisms whose genetic code had been modified by laboratory manipulations, like mice.

4. Concerns would later be raised over the side effects of this hormone in cattle, including an increased risk of mammary infection leading to a need for regular large doses of antibiotics in dairy cattle feed, but this did not stop the use of it in most conventional dairy ranching in the United States.

5. Researchers at the University of Ghent in Belgium discovered the gene transfer mechanism between *Agrobacterium* and plants, which resulted in the development of methods to alter *Agrobacterium* into an efficient delivery system for gene engineering in plants (Schell and Van Montagu 1977; Joos et al. 1983). A team of researchers led by Dr. Mary-Dell Chilton was the first to demonstrate that virulence genes could be removed without adversely affecting the ability of *Agrobacterium* to insert its own DNA into the plant genome (Chilton et al. 1977).

6. For a list of countries growing GE foods see the Genetic Literacy Project, "Where Are GMO Crops and Animals Approved and Banned?" Accessed June 20, 2021. https://geneticliteracyproject.org/gmo-faq/where-are-gmo-crops-and -animals-approved-and-banned/#:~:text=The%20group's%20list%20includes%20 Algeria,and%20importation%20of%20GMO%20crops.

Chapter Three. Ontological Multiplicity & Glyphosate's Safety

1. As it moves from living to nonliving states, we might see glyphosate as acting, in Povinelli's (2016) terms, like a *viral terrorist*. It moves opportunistically through different ontological territories—from chemical to biological forms of existence—refusing to distinguish between life and nonlife in exacting its effects

for better or worse as it goes. In Povinelli's use, viral operators are external to regimes of power that have kept stable orders of colonialism in place, transgressing by not following the ontological and epistemological rules that were set by colonial masters. They provide useful possibilities for Aboriginal engagements with settler colonialism that might be understood as operating through tactics forged in geological modes of power (Povinelli 2016, 2020). The viral aspect of glyphosate's movement is important. Bacteria are self-contained and can live with or without humans, whereas viruses cannot. Viruses survive by replicating themselves in their hosts by way of the host's DNA. The genetic alterations made possible by the bacterial technologies of *A. tumefaciens* in Monsanto laboratories, in fact, worked the same way: by enabling the bacteria to insert itself, as a virus would, into the DNA of the host species, making it a partner to (glyphosate-rich) foods. But viruses are more frightening to humans than bacteria. Bacteria replicate in uncontrollable ways through genetic mutations that make them capable of staying one step ahead of human cellular reproduction, which also makes it possible to target and kill them with the right kind of poisons. Viruses replicate and proliferate *within* hosts, making themselves part of the host's life and survival, until the host dies. To kill a virus, we must think about ways of killing what has become us. In this sense, glyphosate is both a virally partnered thing (when it is used with plants that have been genetically altered with viruses), and something that works like a virus in that it must keep its host alive to replicate itself.

2. The NOAELs produced by Monsanto animal studies were used to determine that glyphosate was safe to humans and that plants with the Bt modification presented no significant risk to animals or humans.

3. Latham's case in point is Agent Orange. We now know the EPA found widespread evidence of fraud and error at the testing companies that led to suggestions that it was safer than was the case (and prompting one testing company executive to say to the EPA regulator, "God, we didn't expect you to look up the raw data!"). Yet the EPA did not shut down their operations or demand new tests. Rather, it strung out the investigation by asking the companies to produce new reports and allowing the test company to send reports without signing them on grounds that they were "still reaching conclusions." Calling its efforts a "salvage operation," Latham notes that the EPA never even looked at 3,000 of the 4,500 studies in question and never questioned the laboratory's obscuring strategies for producing results, which included using historic controls from unrelated studies, elaborate administrative fixes, and accepting fraudulent results. Industrial BioTest (the hired-gun company set up to give the appearance of a disinterested audit of the chemical) alone conducted 35 to 40 percent of all US chemical testing between 1950 and 1990. In 1983 one EPA employee published a report in the *Amicus Journal* (Spring 1983) based on a site visit of the testing facility in which he noted gross negligence in the laboratory's methods. There was apparently evidence of dead animals, leaky controls, and records in such total disarray that specific study data could not be traced. He suspected that many of the results offered

140

to the companies that hired IBT for these reports (i.e., the chemical companies) were falsified. In response, the EPA investigated IBT, including scrutinizing the microfiche records of the results of their studies, and found huge inconsistencies between the study microfiche and the validation reports turned over to the EPA. Latham (2018) notes that this event prompted one company spokesperson to state the passage I've quoted here.

Chapter Five. The Scientific Consensus & the Counterfactual

1. Marie Monique-Robin interviewed Pusztai in her award-winning film (and book) about GE foods, *The World According to Monsanto*.

2. In more detail: "There were ten treatments in the Séralini study: one control group had access to plain water and a standard diet from the closest isogenic non-GE maize; three groups were fed with 11 percent, 22 percent, and 33 percent of GE NK603 maize treated with Roundup in the field; three groups with the same percentages of GE maize but with no Roundup treatment; and three groups with the non-GE maize diet but with the rats' water supplemented with Roundup at 0.50 mg/L, 400 mg/L, or 2,250 mg/L. There were 20 [Sprague Dawley] rats per group—10 males and 10 females, for totals of 20 control rats and 180 treatment rats. The animals were fed for two years, but some animals died before the end of the study. Séralini et al. (2014) measured behavior, appearance, palpable tumors and infections. They also conducted microscopic examinations and biochemical analysis of blood and urine to look for abnormalities. Average tumor incidence reported by Séralini et al. (2014) was comparable to data on untreated Sprague Dawley rats reported by [others]" (NASEM 2016, 189).

3. Many of these are discussed in Perro and Adams (2017), and the studies include Aitbali et al. 2018; Malatesta et al. 2002, 2003, 2008; Vecchio et al. 2004; Richard et al. 2005; Benachour et al. 2007; Gab-Alla et al. 2012; Guilhereme et al. 2012; Shehata et al. 2013; Guyton et al. 2015; Thongprakaisang et al. 2013; Cattani et al. 2014.

4. The list of scientists skeptical of GE who have been attacked is long and includes well-known researchers whom I spoke with at some of the non-GMO workshops: Don Huber, plant pathologist from Purdue; John Fagan, biochemist and molecular biologist from Cornell; Charles Benbrook, agricultural economist formerly at Washington State University; and Michael Hansen, biologist and ecologist with Consumers Union. There is evidence varying from extensive to suggestive that the attacks on Pusztai and Séralini (along with two others we don't include here: Ignacio Chapela and David Quist) were orchestrated by industry. See, for example, the case of Quist/Chapela (Earthsource) (NGIN 2002; Graham 2002). See GMWatch (https://www.gmwatch.org/en/) for regular postings of other examples. See also Matthews 2012; Robinson and Latham 2013; Leoning 2012.

141

5. See also Judy Carman's pages on the Sustainable Pulse website, https://GMOjudycarman.org.

Chapter Six. Consensuses, Academic Capitalism & the Swirl

1. Interestingly enough, this same magazine published an article six years earlier proclaiming the health harms from glyphosate (Gammon 2009).

2. She reminds us that historically there have never been clear lines dividing the world of theoretical (disinterested or basic) science from applied (industry) science. In fact, the idea that industry-driven science is not "basic" had to be carved out from research milieus where profitable and impactful ideas ("golden eggs," she calls them) were always being produced from theoretical work that was also quite basic (consider Pasteur or Curie, she says).

3. I am not suggesting that normative science is ever really in the business of providing judgment on the basis of the facts it produces; scientific studies offer facts that people use to make claims about how harmful something is or what people should do about it (Pinch and Bijker 2012).

Chapter Seven. Glyphosate Becomes an Activist

1. See "Cal-EPA (OEHHA)—Glyphosate Hearing 6-7-17," YouTube video, accessed January 21, 2018, https://www.youtube.com/watch?v=zBU_ -HeWfNI&feature=em-share_video_user.

2. Honeycutt's numbers were as follows:

TABLE 7.1. TYPICAL CHILD'S CONSUMPTION OF GLYPHOSATE (2017)

Food	Glyphosate Detected ppb (10.9 mcg p/b)	Proposed California NSRL (mcg/kg)
Oatmeal	70.9	1100
Hummus and pita	1640	5100
Milk	30.7	allowable level not yet set
Corn chips	56	2547
Berries	0	22.6
Water	257	700
Eggs	14.3	4.2
Orange juice	2.2	42.2
Toast and jam	187	1100
Pasta	226.3	560

SOURCE: "CAL-EPA (OEHHA)—GLYPHOSATE HEARING 6-7-17."

3. These documents may be accessed at US Right to Know, "Roundup (Glyphosate) Cancer Cases: Key Documents and Analysis," web page, accessed July 21, 2021, https://usrtk.org/monsanto-papers/.

Chapter Eight. Chemicals as Agents of Care

1. A useful primer can be found in an interview with Puig de la Bellacasa and Dimitris Papadopoulos as part of *The Swamp School*, the Lithuanian pavilion presented at the 2018 Venice International Architecture Biennale. A recording is available online: "Decenter the Human: Interview with D. Papadopoulos and M. P. de la Bellacasa," YouTube video, accessed March 16, 2019, https://www .youtube.com/watch?v=YzAO-RqiuTs.

Ackerman-Leist, Philip. 2017. *A Precautionary Tale: How One Small Town Banned Pesticides, Preserved Its Food Heritage, and Inspired a Movement*. White River Junction, VT: Chelsea Green.

Adamovski, Ezequiel. 2014. "Andres Carrasco vs. Monsanto." *Telesur TV*, October 1, 2014. https://www.telesurtv.net/english/opinion/Andres-Carrasco-vs-Monsanto-20141010-090.html.

Adams, Axel, and Roy Gerona. 2016. "Biomonitoring of Glyphosate across the United States in Urine and Tap Water Using High-Fidelity LC-MS/MS Method." Poster presented at the University of California, San Francisco, May 25, 2016. Accessible at The Detox Project. "UCSF Presentation Reveals Glyphosate Contamination in People across America." https://detoxproject.org/13212-/.

Adams, Mike. 2017. "Heartbreaking Letter from Dying EPA Scientist Begs Monsanto 'Moles' inside the Agency to Stop Lying about Dangers of RoundUp (Glyphosate)." US Right to Know, March 15, 2017. https://www.naturalnews.com/2017-03-15-heartbreaking-letter-from-dying-epa-scientist-begs-monsanto-moles-inside-the-agency-to-stop-lying-about-dangers-of-roundup-glyphosate.html.

Adams, Vincanne, ed. 2016. "Metrics of the Global Sovereign." In *Metrics: What Counts in Global Health*, edited by Vincanne Adams, 19–54. Durham, NC: Duke University Press.

Agard-Jones, Vanessa. 2013. "Bodies in the System." *Small Axe* 17 (3): 182–92.

Agard-Jones, Vanessa. 2014. "Spray." *Somatosphere*, May 17, 2014. http://somatosphere.net/2014/spray.html/.

Aitbali, Yassine, Saadia Ba-M'hamed, Najoua Elhidar, Ahmed Nafis, Nabila Soraa, and Mohammed Bennis. 2018. "Glyphosate-Based Herbicide Exposure Affects Gut Microbiota, Anxiety and Depression-Like Behaviors in Mice." *Neurotoxicology and Teratology* 67: 44–49.

Andersson, Hans Christer, Detlef Bartsch, Hans-Joerg Buhk, Howard Davies, Marc De Loose, Michael Gasson, Niels Hendriksen, et al. 2005. "Opinion of the Scientific Panel on Genetically Modified Organisms [GMO] Related to the Notification for the Placing on the Market of Insect Resistant Genetically Modified Maize Bt11, for Cultivation, Feed and Industrial Processing." *European Food Safety Authority Journal* 3 (5): 213.

Antonetta, Susanne. 2001. *Body Toxic: An Environmental Memoir*. Washington, DC: Counterpoint.

Antoniou, M. N., Armel Nicolas, Robin Mesnage, Martini Biserni, Francesco V. Rao, and Cristina Vasquez Martin. 2019. "Glyphosate Does Not Substitute for Glycine in Proteins of Actively Dividing Mammalian Cells." *BMC Research Notes* 12: 494.

Appadurai, Arjun, ed. 1986. *The Social Life of Things: Commodities in Cultural Perspective*. Cambridge: Cambridge University Press.

Arcuri, Alessandra, and Yogi Hendlin. 2019. "The Chemical Anthropocene: Glyphosate as a Case Study of Pesticide Exposures." *King's Law Journal* 30 (2): 234–53.

Arcuri, Alessandra, and Yogi Hendlin. 2020. "Introduction to the Symposium on the Science and Politics of Glyphosate." *European Journal of Risk Regulation* 11 (3): 411–21.

Avila-Vazquez, Medardo, Eduardo Maturano, Agustina Etchegoyen, Flavia Sylvina Difilippo, and Bryan Maclean. 2017. "Association between Cancer and Environmental Exposure to Glyphosate." *International Journal of Clinical Medicine* 8 (2): 73–85. https://www.scirp.org/journal/paperinformation .aspx?paperid=74222#:~:text=This%20research%20detected%20an%20 urban,to%20make%20direct%20causal%20assertions.

Balayannis, Angeliki. 2020. "Toxic Sights: The Spectacle of Hazardous Waste Removal." *Environment and Planning D: Society and Space* 38 (4): 772–90.

Balayannis, Angeliki, and Emma Garnett. 2020. "Chemical Kinship: Interdisciplinary Experiments with Pollution." *Catalyst* 6 (1). www.https://doi.org/10 .28968/cftt.v6i1.33524.

Ball, D. A., C. Rainbolt, D. C. Thill, and J. P. Yenish. 2003. "Weed Management Strategies for CLEARFIELD* Wheat Systems across PNW Precipitation Zones." Paper presented at the Sixth Annual Northwest Direct Seed Cropping Systems Conference and Trade Show, Pasco, Washington, January 8–10, 2003. http://pnwsteep.wsu.edu/directseed/conf2k3/dsc3ball3.htm.

Barad, Karen. 2010. "Quantum Entanglements and Hauntological Relations of Inheritance: Dis/continuities, SpaceTime Enfoldings, and Justice-to-Come." *Derrida Today* 3 (2): 240–68.

Baum Hedlund. 2022. "Where Is Glyphosate Banned?" Website, updated March 2022. https://www.baumhedlundlaw.com/toxic-tort-law/monsanto -roundup-lawsuit/where-is-glyphosate-banned-/.

Baum Hedlund Aristei & Goldman PC. 2017. "EPA Officials Helped Monsanto Kill Glyphosate Study." Blog, March 14, 2017. https://www.baumhedlundlaw .com/blog/2017/march/epa-official-helped-monsanto-kill-glyphosate-stu/.

Bayer Group. 2021. "Spray Smarter, Know Your Options: The History of Roundup." Crop Science/Canada website, updated November 3, 2021. https://www.cropscience.bayer.ca/Products/Herbicides/Roundup.

Benachour, N., H. Sipahutar, S. Moslemi, C. Gasnier, C. Travert, and G. E. Séralini. 2007. "Time- and Dose-Dependent Effects of Roundup on Human Embryonic and Placental Cells." *Archives of Environmental Contamination and Toxicology* 53 (1): 126–33.

Benbrook, Charles M. 2016. "Trends in Glyphosate Herbicide Use in the United States and Globally." *Environmental Sciences Europe* 28 (1): 3. https://www.ncbi .nlm.nih.gov/pmc/articles/PMC5044953/.

Benbrook, Charles, M. 2019. "How Did the US EPA and IARC Reach Diametrically Opposed Conclusions on the Genotixicity of Glyphosate-Based Herbicides?" *Environmental Sciences Europe* 31 (2). https://d-nb.info/1178263355/34.

Bero, Lisa. 2018. "Lisa Bero: More Journals Should Have Conflict of Interest Policies as Strict as Cochrane." *BMJ Opinion*, November 12, 2018. https://blogs .bmj.com/bmj/2018/11/12/lisa-bero-more-journals-should-have-conflict-of -interest-policies-as-strict-as-cochrane/.

Blanchette, Alex. 2020. *Porkopolis: American Animality, Standardized Life, and the Factory Farm.* Durham, NC: Duke University Press.

Blancke, Stefaan. 2015. "Why People Oppose GMOs Even Though Science Says They Are Safe." *Scientific American*, August 18, 2015. https://www.scientificamerican.com/article/why-people-oppose-gmos-even-though-science-says-they-are-safe/.

Bo, Ernesto Dal. 2006. "Regulatory Capture: A Review." *Oxford Review of Economic Policy* 22 (2): 203–25.

Boudia, Soraya. 2014. "Managing Scientific and Political Uncertainty: Environmental Risk Assessment in a Historical Perspective." In Boudia and Jas, *Powerless Science? Science and Politics in a Toxic World*, vol. 2, chap. 4, loc. 2065.

Boudia, Soraya, and Nathalie Jas, eds. 2014a. "Introduction: The Greatness and Misery of Science in a Toxic World." In Boudia and Jas, *Powerless Science? Science and Politics in a Toxic World*, vol. 2, introduction, loc. 109.

Boudia, Soraya, and Nathalie Jas, eds. 2014b. *Powerless Science? Science and Politics in a Toxic World.* Vol. 2. New York: Berghahn Books. Kindle edition.

Boudia, Soraya, and Nathalie Jas, eds. 2015. *Toxicants, Health and Regulation since 1945.* New York: Routledge.

Bravo, Kristina. 2014. "Here's How the World's Largest Biotech Company Came to Be." TakePart, March 27, 2014. https://www.thelibertybeacon.com/heres-how-the-worlds-largest-biotech-company-came-to-be/.

Callon, Michael, Cecile Meadel, and Vololona Raherabisoa. 2002. "The Economy of Qualities." *Economy and Society* 31 (2): 194–217.

Carman, Judy A., Howard R. Vlieger, Larry J. Ver Steeg, Verlyn E. Sneller, Garth W. Robinson, Catherine A. Clinch-Jones, Julie I. Haynes, and John W. Edwards. 2013. "A Long-Term Toxicology Study on Pigs Fed a Combined Genetically Modified (GM) Soy and GM Maize Diet." *Journal of Organic Systems* 8 (1): 38–54.

Carmody, Rachel N., Howard R. Vlieger, Larry J. Ver Steeg, Verlyn E. Sneller, Garth W. Robinson, Catherine A. Clinch-Jones, Lulie I. Haynes, et al. 2015. "Diet Dominates Host Genotype in Shaping the Murine Gut Microbiota Cell." *Host and Microbe* 17 (1): 72–84.

Carson, Rachel. 1962. *Silent Spring.* New York: Houghton Mifflin.

Caruso, Denise. 2007. "A Challenge to Gene Theory, a Tougher Look at Biotech." *New York Times*, July 1, 2007. https://www.nytimes.com/2007/07/01/business/yourmoney/01frame.html.

Cattani, Diane, Vera Lúcia de Liz Oliveira Cavalli, Carla Elise Heinz Rieg, Juliana Tonietto Domingues, Tharine Dal-Cim, Carla Inês Tasca, Fátima Regina Mena Berreto Silva, and Ariane Zamoner. 2014. "Mechanisms Underlying the Neurotoxicity Induced by Glyphosate-Based Herbicide in Immature Rat Hippocampus: Involvement of Glutamate Excitotoxicity." *Toxicology* 320: 34–45.

Cavagna, Andrea, Alessio Cimarelli, Irene Giardina, and Massimiliano Viale. 2010. "Scale-Free Correlations in Starling Flocks." *PNAS* 107 (26): 11865–70.

Chen, Mel Y. 2011. "Toxic Animacies, Intimate Affections." *GQL: A Journal of Lesbian and Gay Studies* 17 (2–3): 265–86.

147

Chilton, Mary-Dell, Martin H. Drummond, Donald J. Merlo, Daniela Sciaky, Alice L. Montoya, Milton P. Gordon, and Eugene W. Nester. 1977 "Stable Incorporation of Plasmid DNA into Higher Plant Cells: The Molecular Basis of Crown Gall Tumorigenesis." *Cell* 11 (2): 263–71. https://doi.org/10.1016/0092-8674(77)90043-5.

Chiu, Kellia, Quinn Grundy, and Lisa Bero. 2017. "'Spin' in Published Medical Literature: A Methodological and Systematic Review." *PLOS Biology* 15 (9). https://doi.org/10.1371/journal.pbio.2002173.

Church, Norman J. 2005. "Why Our Food Is So Dependent on Oil." *Resilience*, April 1, 2005. https://www.resilience.org/stories/2005/0-40-1/why-our-food-so-dependent-oil/.

Clement, Roland C. 1972. "The Pesticides Controversy." *Boston College Environmental Affairs Law Review* 2 (3): 445–68. http://lawdigitalcommons.bc.edu/ealr/vol2/iss3/1.

Conis, Elena. 2010. "Debating the Health Effects of DDT: Thomas Jukes, Charles Wurster, and the Fate of an Environmental Pollutant." *Public Health Reports* 125 (2): 337–42. https://www.ncbi.nlm.nih.gov/pmc/articles/PMC2821864.

Conis, Elena. 2016. "DDT Disbelievers: Health and the New Economic Poisons in Georgia after World War II." *Southern Spaces*, October 28, 2016. https://southernspaces.org/2016/ddt-disbelievers-health-and-new-economic-poisons-georgia-after-world-war-ii/.

Conroy, Joan. 2018. "Science Advocate Alison Van Eenannaam." Alliance for Science, October 9, 2018. https://allianceforscience.cornell.edu/blog/2018/10/profile-science-advocate-alison-van-eenennaam/.

Cook, Joan. 1990. "Howard A. Schneiderman, 63, Expert on Altering Plant's Genes." Obituary, *New York Times*, December 7, 1990. https://www.nytimes.com/1990/12/07/obituaries/howard-a-schneiderman-63-expert-on-altering-plants-genes.html.

Copley, Marion. 2013. "Marion Copley to Jess Rowland, March 4, 2013." Document 1411-, Roundup Products Liability Litigation, Industry Documents Library, University of California, San Francisco.

Council on Scientific Affairs. 1985. "Saccharin: Review of Safety Issues." *JAMA* 254 (18): 2622–24.

Creager, Angela N. H., and Jean-Paul Gaudillière, eds. 2021. *Risk on the Table: Food Production, Health, and the Environment*. New York: Berghahn Books.

Davis, Frederick Rowe. 2014. *Banned: A History of Pesticides and the Science of Toxicology*. New Haven, CT: Yale University Press.

Dawson, Bethany. 2021. "More Than 200 Dead Birds Fell from the Sky, Hitting Pedestrians and Vehicles in Spain, Reports Say." *Insider*, December 5, 2021. https://www.insider.com/spain-dead-birds-rain-from-the-sky-hitting-pedestrians-vehicles-2021-2.

Deleuze, Gilles, and Felix Guattari. 1987. *A Thousand Plateaus: Capitalism and Schizophrenia*. Minneapolis: University of Minnesota Press.

Delson, Sam. 2017. "Glyphosate to be Added to Proposition 65 List of Chemicals." OEHHA, March 28, 2017. https://oehha.ca.gov/public-information/press -release/press-release-proposition-65/glyphosate-be-added-proposition-65.

Dill, G. M., R. D. Sammons, P. C. Feng, F. Kohn, K. Kretzmer, A. Mehrsheikh, M. Bleeke, et al. 2010. "Glyphosate: Discovery, Development, Applications, and Properties." In *Glyphosate Resistance in Crops and Weeds*, edited by V. K. Nandula, 1–33. Hoboken, NJ: Wiley.

Domingo, José L., and Jordi Giné Bordonaba. 2011. "A Literature Review on the Safety Assessment of Genetically Modified Plants." *Environment International* 37: 734–42.

Dowd-Uribe, Brian. 2014. "Engineering Yields and Inequality? How Institutions and Agro-Ecology Shape Bt Cotton Outcomes in Burkina Faso." *Geoforum* 53 (May): 161–71.

Druker, Steven M. 2015. *Altered Genes, Twisted Truth: How the Venture to Genetically Engineer Our Food Has Subverted Science, Corrupted Government, and Systematically Deceived the Public*. White River Junction, VT: Clear River Press.

Dumit, Joseph. 2012. *Drugs for Life: How Pharmaceutical Companies Define Our Health*. Durham, NC: Duke University Press.

EarthSource. 2017. "Myth: Independent Studies Confirm That GM Foods and Crops Are Safe." Earth Open Source. http://earthopensource.org /gmomythsandtruths/sample-page/2-science-regulation/22-myth -independent-studies-confirm-gm-foods-crops-safe.

Edelman, Lee. 2004. *No Future: Queer Theory and the Death Drive*. Durham, NC: Duke University Press.

Edwards, Greg. 2021. "Investors Sue Bayer over Monsanto Purchase, Decline in Stock." *St. Louis Business Journal*, January 25, 2021. https://www.bizjournals .com/stlouis/news/2021/01/25/investors-sue-bayer-over-monsanto-purchase -declin.html.

EFSA (European Food Safety Authority). 2012. "Review of the Séralini et al. (2012) Publication on a 2-Year Rodent Feeding Study with Glyphosate Formulations and GM Maize NK603 as Published Online on 19 September 2012 in *Food and Chemical Toxicology*." *EFSA Journal* 10 (11): 2986.

EFSA (European Food Safety Authority). 2015. "Conclusion on the Peer Review of the Pesticide Risk Assessment of the Active Substance Glyphosate." *EFSA Journal* 13 (11): 4302. https://www.efsa.europa.eu/en/efsajournal/pub/4302.

EFSA Scientific Committee. 2011. "Statistical Significance and Biological Relevance." *EFSA Journal* 9 (9): 2372.

Egelko, Bob. 2018. "Benicia Man Dying of Cancer Testifies in Roundup Case." *San Francisco Chronicle*, July 24, 2018. https://www.sfchronicle.com/bayarea /article/Benicia-man-dying-of-cancer-testifies-in-Roundup-13098953.php.

Egelko, Bob. 2019. "Monsanto's Roundup Found by Jury to Be Likely Cause of Cancer in Second Bay Area Man." *San Francisco Chronicle*, March 19, 2019. https://www.sfchronicle.com/bayarea/article/Benicia-man-dying-of-cancer -testifies-in-Roundup-13098953.php.

El-Shamei, Z. S., A. A. Gab-Alla, A. A. Shatta, E. A. Moussa, and A. M. Rayan. 2012. "Histopathological Changes in Some Organs of Male Rats Fed on Genetically Modified Corn (Ajeeb YG)." *Journal of American Science* 8 (10): 684–96.

Eng, Monica. 2016. "Why Didn't an Illinois Professor Have to Disclose GMO Funding?" *WBEZ Chicago*, March 15, 2016. https://www.wbez.org/stories/why -didnt-an-illinois-professor-have-to-disclose-gmo-funding/eb99bdd26-83d -41089-528-de1375c3e9fb.

EPA (Environmental Protection Agency). 2020. "EPA Finalizes Glyphosate Miti- gation." News release, January 30, 2020. https://www.epa.gov/pesticides/epa -finalizes-glyphosate-mitigation.

Eskenazi, Brenda A., A. Bradman, and R. Castorina. 1999. "Exposures of Children to Organophosphate Pesticides and Their Potential Adverse Health Effects." *Environmental Health Perspectives* 107, supplement 3 (June): 409–19.

European Commission. 2010. "Commission Recommendations of the 13, July 2010 on Guidelines for the Development of National Co- existence Measures to Avoid the Unintended Presence of GMOs in Conventional and Organic Crops 2010." *Official Journal of the European Union.* https://food.ec.europa.eu/system/files/2016-10/ plant_gmo-agriculture_coexistence-new_recommendation_en.pdf.

Ewen, S. W., and A. Pusztai. 1999. "Effect of Diets Containing Genetically Modi- fied Potatoes Expressing *Galanthus nivalis* Lectin on Rat Small Intestine." *Lancet* 354 (9187): 1353–54.

Fagan, John, Terje Traavik, and Thomas Bohn. 2015. "The Séralini affair: De- generation of Science to Re-Science?" *Environmental Sciences Europe* 27 (19). https://www.aminer.cn/pub/56d9286ddabfae2eeec93622/the-seralini-affair -degeneration-of-science-to-re-science.

Fasano, Allesio. 2012. "Leaky Gut and Autoimmune Diseases." *Clinical Reviews in Allergy and Immunology* 42 (1): 71–78.

Finlay, Brett, and Marie-Claire Arrieta. 2016. *Let Them Eat Dirt: Saving Your Child from an Oversanitized World.* New York: Algonquin Books.

Fortun, Kim. 2012a. "Ethnography in Late Industrialism." *Cultural Anthropology* 27 (3): 446–64.

Fortun, Kim. 2012b. "Biopolitics and the Informating of Environmentalism." In *Lively Capital: Biotechnologies, Ethics, and Governance in Global Markets*, edited by Kaushik Rajan, 306–26. Durham, NC: Duke University Press.

Fox-Keller, Evelyn. 2009. *The Century of the Gene.* Cambridge, MA: Harvard University Press.

Franklin, Sarah. 2007. *Dolly Mixtures: The Remaking of Genealogy.* Durham, NC: Duke University Press.

Frickel, Scott, and Michelle Edwards. 2014. "Untangling Ignorance in Environ- mental Risk Assessment." In Boudia and Jas, *Powerless Science? Science and Politics in a Toxic World*, vol. 2, chap. 10, loc. 4646.

Frickel, Scott, and Kelly Moore, eds. 2015. *The New Political Sociology of Science: Institutional Networks and Power*. Madison: University of Wisconsin Press.

Gab-Alla, El-Shamei, Moussa Adel Shatta, and Ahmed Mohamed Rayan. 2012. "Morphological and Biochemical Changes in Male Rats Fed on Genetically Modified Corn (Ajeeb YG)." *Journal of American Science* 8 (9): 1117–23.

Gaber, Nadia. 2019. "Life after Water: Detroit, Flint and the Postindustrial Politics of Health." PhD diss., University of California, San Francisco.

Gabriel, Joseph. 2016. "Pharmaceutical Patenting and the Transformation of American Medical Ethics." *British Journal of the History of Science* 49 (4): 577–600.

Gammon, Crystal. 2009. "Weed-Whacking Herbicide Proves Deadly to Human Cells." *Scientific American*, June 23, 2009. https://www.scientificamerican.com /article/weed-whacking-herbicide-p/#:~:text=Used%20in%20yards%2C%20 farms%20and,placental%20and%20umbilical%20cord%20cells.

Gaudillière, Jean-Paul. 2017. *The Development of Scientific Marketing in the Twentieth Century: Research for Sales in the Pharmaceutical Industry*. New York: Routledge.

Gemmill, Alison, Robert B. Gunier, Asa Gradman, Brenda Eskenazi, and Kim G. Harley. 2014. "Residential Proximity to Methyl Bromide Use and Birth Outcomes in an Agricultural Population in California." *Environmental Health Perspectives* 121 (6): 737–43.

Genetic Literacy Project. n.d. "Where Are GMO Crops and Animals Approved and Banned?" Accessed June 25, 2022. https://geneticliteracyproject.org /gmo-faq/where-are-gmo-crops-and-animals-approved-and-banned.

Gilbert, Lawrence I. 1994. "Howard Schneiderman." In *Biographical Memoirs 63*, 481–502. Washington, DC: National Academies Press. https://www.nap.edu /read/4560/chapter/22#493.

Gilham, Carey. 2017. *Whitewash: The Story of a Weed Killer, Cancer, and the Corruption of Science*. Washington, DC: Island Press.

Gilham, Carey. 2018. "I Won a Historic Lawsuit, but May Not Live to Get the Money." *Time*, November 21, 2018. https://time.com/5460793/dewayne-lee -johnson-monsanto-lawsuit/.

GMO Answers. 2015. "IARC's Classification of Glyphosate—What Does It Mean for You?" Accessed July 12, 2021. https://gmoanswers.com/iarc%25E2%2580%2599s -classification-glyphosate-%25E2%2580%2593-what-does-it-mean-you.

Goldberg, Ray A. 2018. *Food Citizenship: Food System Advocates in an Era of Mistrust* Oxford: Oxford University Press.

Gould, Fred. 2018. "Pesticide Resistance Arms Race." Interview by Tracey Peake. Genetic Engineering and Society Center, North Carolina State University, June 29, 2018. https://research.ncsu.edu/ges/2018/06/pesticide-resistance -arms-race/.

Graf, Jeorg. 2011. "Shifting Paradigm on *Bacillus thuringiensis* Toxin and a Natural Model for *Enterococcus faecalis* Septicemia." *mBio* 2 (4): e00161–11.

Graham, Sarah. 2002. "Journal Retracts Support for Claims of Invasive GM Corn." *Scientific American*, April 8, 2002. https://www.scientificamerican .com/article/journal-retracts-support/.

151

Gray, Brice. 2016. "From Sweeteners to Seeds: A Timeline of Monsanto Evolution." *St. Louis Post-Dispatch*, September 14, 2016. https://www.stltoday.com/business/local/from-sweeteners-to-seeds-a-timeline-of-monsantos-evolution/article_bae32ab7-e2ec-506a-a9cd-16b9753b909b.html.

Guilhereme, S., O Gaivão, M. A. Santos, and M. Pacheco. 2012. "DNA Damage in Fish (*Anguilla anguilla*) Exposed to a Glyphosate-Based Herbicide—Elucidation of Organ-Specificity and the Role of Oxidative Stress." *Mutation Research* 743 (1–2): 1–9.

Guillette, E. A., M. M. Meza, M. G. Aguilar, A. D. Soto, and I. E. Garcia. 1998. "An Anthropological Approach to the Evaluation of Preschool Children Exposed to Pesticides in Mexico." *Environmental Health Perspectives* 106 (6): 347–53.

Guthman, Julie. 2011. *Weighing In: Obesity, Food Justice, and the Limits of Capitalism*. Berkeley: University of California Press.

Guthman, Julie. 2014. *Agrarian Dreams: The Paradox of Organic Farming in California*. Berkeley: University of California Press.

Guthman, Julie. 2019. *Wilted: Pathogens, Chemicals, and the Fragile Future of the Strawberry Industry*. Berkeley: University of California Press.

Guyton, Kathryn Z., Dana Loomis, Yann Grosse, Fatiha El Ghissassi, Lamia Benbrahim-Tallaa, Neela Guha, Chiara Scoccianti, Heidi Mattock, and Kurt Straif. 2015. "Carcinogenicity of Tetrachlorvinphos, Parathion, Malathion, Diazinon, and Glyphosate." *Lancet: Oncology* 16 (5): 490–91.

Hals, Tom, and Tina Bellon. 2021. "Bayer Reaches $2 Billion Deal over Future Roundup Cancer Claims." *Reuters*, February 3, 2021. https://www.reuters.com/article/us-bayer-glyphosate/bayer-reaches-2-billion-deal-over-future-roundup-cancer-claims-idUSKBN2A32MX.

Halter, S. 2007. "A Brief History of Roundup." In *Proceedings of the First International Symposium on Glyphosate, Agronomical Sciences College of the University of the State of São Paulo, São Paulo, Brazil, October 15–19, 2007*.

Harrison, Jill Lindsay. 2006. "'Accidents' and Invisibilities: Scaled Discourse and the Naturalization of Regulatory Neglect in California's Pesticide Drift Conflict." *Political Geography* 25 (5): 506–29.

Hayes, Tyrone. B., Atif Collins, Melissa Lee, Magdalena Mendoza, Nigel Noriega, A. Ali Stuart, and Aaron Vonk. 2002. "Hermaphroditic, Demasculinized Frogs after Exposure to the Herbicide Atrazine at Low Ecologically Relevant Doses." *PNAS* 99 (8): 5476–80.

Hayes, Tyrone B., and Martin Hanson. 2017. "From Silent Spring to Silent Night: Agrochemicals and the Anthropocene." *Elementa: Science of the Anthropocene* 5: 57.

Heimbuch, Jaymi. 2019. "The Incredible Science behind Starling Murmurations." Treehugger, December 27, 2019. https://www.treehugger.com/the-incredible-science-behind-starling-murmurations-4863751.

Hendlin, Yogi. 2019. Presentation to the Environmental Health Institute, University of Southern California, San Francisco, December 10, 2019.

Hendlin, Yogi. 2021. "Surveying the Chemical Anthropocene." *Environment and Society: Advances in Research* 12 (1): 181–202.

Hessler, Uwe. 2020. "What's Driving Europe's Stance on Glyphosate?" DW Akademie, June 5, 2020. https://p.dw.com/p/3eGKA.

Hetherington, Kregg. 2013. "Beans before the Law: Knowledge Practices, Responsibility, and the Paraguayan Soy Bean Boom." *Cultural Anthropology* 28 (1): 65–85.

Hilbeck, Angelika, Rosa Binimelis, Nicolas Defarge, Ricarda Steinbrecher, András Székács, Fern Wickson, Michael Antoniou, et al. 2015. "No Scientific Consensus on GMO Safety." *Environmental Sciences Europe* 27 (4). https://enveurope.springeropen.com/articles/10.1186/s12302-014-0034-1.

Hilbeck, Angelika, Hartmut Meyer, Brian Wynne, and Erik Millstone. 2020. "GMO Regulations and Their Interpretation: How EFSA's Guidance on Risk Assessments of GMOs Is Bound to Fail." *Environmental Sciences Europe* 32 (54). www.https://doi.org/10.1186/s123020-200-03256-.

Höfte, Herman, Henri de Greve, Jef Seurinck, Stefan Jansens, Jacques Mahillon, Christophe Ampe, Joel Vandekerckhove, Hilde Vanderbruggen, Marc Van Montagu, Marc Zabeau, and Mark Vaeck. 1986. "Structural and Functional Analysis of a Cloned Delta Endotoxin of *Bacillus thuringiensis berliner 1715*." European Journal of Biochemistry. 161 (2) (Dec. 1): 273–80. doi: 10.1111/j.1432-1033.1986.tb10443.x. PMID: 3023091.

Holmes, Seth. 2013. *Fresh Fruit, Broken Bodies: Migrant Farmworkers in the United States.* Berkeley: University of California Press.

Honeycutt, Zen, and Henry Rowlands. 2014. "Glyphosate Testing Full Report: Findings in American Mothers' Breast Milk, Urine and Water." Moms across America, April 7, 2014. http://www.momsacrossamerica.com/glyphosate_testing_results.

Hoover, Elizabeth. 2017. *The River Is in Us: Fighting Toxics in a Mohawk Community.* Minneapolis: University of Minnesota Press.

Horton, Richard. 1999. "Genetically Modified Foods: 'Absurd' Concern or Welcome Dialogue?" *Lancet* 354: 1314–15.

Hough, Peter. 1998. *The Global Politics of Pesticides: Forging Consensus from Conflicting Interests.* New York: Routledge.

Huber, M., E. Rambialkowska, D. Srednicka, S. Bugel, and L. P. L van de Vijner. 2011. "Organic Food and Impact on Health: Assessing the Status Quo and Prospects of Research." *NJAS: Wageningan Journal of Life Sciences* 58: 103–9.

IARC (International Agency for Research on Cancer). 2015. "Evaluation of Five Organophosphate Insecticides and Herbicides." IARC Monographs. World Health Organization, March 20, 2015. https://www.iarc.fr/en/media-centre/iarcnews/pdf/MonographVolume112.pdf.

Jain, S. Lochlann. 2006. *Injury: The Politics of Product Design and Safety Law in the United States.* Princeton, NJ: Princeton University Press.

Jain, S. Lochlann. 2013. *Malignant: How Cancer Becomes Us.* Berkeley: University of California Press.

Jas, Nathalie, and Soraya Boudia. 2013. *Toxicants, Health and Regulation since 1945.* New York: Routledge.

153

Jeschke, Mark. n.d. "Weed Management in the Era of Glyphosate Resistance." *Crop Insights*, accessed January 13, 2021. https://www.pioneer.com/us /agronomy/weed_mgmt_era_glyphosate_resistance.html.

Joos, H., B. Timmerman, M. V. Montagu, and J. Schell. 1983. "Genetic Analysis of Transfer and Stabilization of *Agrobacterium* DNA in Plant Cells." *EMBO Journal* 2 (12): 2151–60.

Jukes, Thomas H. 1971. "DDT, Human Health and the Environment." *Boston College Environmental Affairs Law Review* 3 (1): 534–63. http://lawdigitalcommons .bc.edu/ealr/vol1/iss3/4.

Kannisery, Ramdas, Biwek Gairhe, Davie Kadyampakeni, Ozgur Batuman, and Fernando Alferez. 2019. "Glyphosate: Its Environmental Persistence and Impact on Crop Nutrition." *Plants* (11): 499.

Kaufman, Sharon R. 2015. *Ordinary Medicine: Extraordinary Treatments, Longer Lives and Where to Draw the Line*. Durham, NC: Duke University Press.

Kaufman, Sharon R. 2019. "Whither Physician Talk and Medicine's Tools?" *Cambridge Quarterly of Healthcare Ethics* 28 (3): 405–9.

Kelland, Kate. 2017. "In Glyphosate Review, WHO Cancer Agency Edited Out 'Non-carcinogenic' Findings." *Reuters Investigates*, October 19, 2017. https:// www.reuters.com/investigates/special-report/who-iarc-glyphosate/.

Kenner, Alison. 2018. *Breath Taking: Asthma Care in a Time of Climate Change*. Minneapolis: University of Minnesota Press.

Kessler, David A. 2010. *The End of Overeating: Taking Control of the Insatiable American Appetite*. New York: Rodale.

Kevles, Daniel J. 2002. *A History of Patenting Life in the United States with Comparative Attention to Europe and Canada*. Luxembourg: Office for Official Publications of the European Communities.

Kirksey, Eben. 2020. "Chemosociality in Multispecies Worlds: Endangered Frogs and Toxic Possibilities in Sydney." *Environmental Humanities* 12 (1): 23–50.

Klein, Kendra, and Kari Hamerschlag. 2016. "Scientific American Science Panel May Get Lost in Translation." Friends of the Earth, March 30, 2016. https:// foe.org/blog/scientific-american-science-panel-may-get-lost-translation/.

Kloppenberg, Jack Ralph. 2005. *First the Seed: The Political Economy of Plant Biotechnology*. 2nd ed. Madison: University of Wisconsin Press.

Koester, Vera. 2017. "What Is Glyphosate?" *Chemistry Views*, September 5, 2017. https://www.chemistryviews.org/details/education/5132411/What_is _Glyphosate.html.

Komives, Tamas, and Peter Schroder. 2016. "On Glyphosate." *Ecocycles* 2 (2): 1–8.

Krimsky, Sheldon. 2014. "Low-Dose Toxicology: Narratives from the Science-Transcience Interface." In Boudia and Jas, *Powerless Science? Science and Politics in a Toxic World*, vol. 2, chap. 11, loc. 5037.

Lamoreaux, Janelle. 2016. "What If the Environment Is a Person? Lineages of Epigenetic Science in a Toxic China." *Cultural Anthropology* 31 (2): 188–214.

Lancaster, Roger. 2011. *Sex Panic and the Punitive State*. Berkeley: University of California Press.

Landecker, Hannah. 2010. *Culturing Life: How Cells Became Technologies*. Cambridge, MA: Harvard University Press.

Landecker, Hannah. 2011. "Food as Exposure: Nutritional Epigenetics and the Molecular Politics of Eating." *Biosocieties* 6 (2): 167–94.

Landecker, Hannah. 2019. "A Metabolic History of Manufacturing Waste: Food Commodities and Their Outsides." *Food, Culture and Society* 22 (5): 530–47.

Landrigan, Philip J., and Charles Benbrook. 2015. "GMOs, Herbicides, and Public Health." *New England Journal of Medicine* 373: 693–95.

Ladrum, Asheley R., William K. Hallman, and Kathleen Hall Jamieson. 2019. "Examining the Impact of Expert Voices: Communicating the Scientific Consensus on Genetically Modified Organisms." *Environmental Communication* 13 (1): 51–70.

Latham, Jonathan. 2018. "Unsealing the Science: What the Public Can Learn from Internal Chemical Industry Documents." Lecture presented at the University of California, San Francisco, September 13, 2018.

Latour, Bruno. 1988. *Science in Action: How to Follow Scientists and Engineers through Society*. Cambridge, MA: Harvard University Press.

Latour, Bruno. 2007. *Reassembling the Social: An Introduction to Actor-Network-Theory*. Oxford: Oxford University Press.

Latour, Bruno. 2005. *Reassembling the Social: An Introduction to Actor Network Theory*. Oxford: Oxford University Press.

Latour, Bruno, and Steve Woolgar. 1979. *Laboratory Life: The Social Construction of Scientific Facts*. Princeton, NJ: Princeton University Press.

Lee, Li-Young. 1986. "From Blossoms." In *Rose*. Rochester, NY: BOA Editions.

Leonard-Barton, Dorothy, and Gary P. Pisano. 1990. "Monsanto's March into Biotechnology." Harvard Business Review Case Studies. Cambridge, MA: Harvard Business School.

Leoning, Ulrich E. 2012. "A Challenge to Scientific Integrity: A Critique of the Critics of the GMO Rat Study Conducted by Gilles-Eric Séralini et al. (2012)." *Environmental Sciences Europe* 27 (13). https://enveurope.springeropen.com/articles/10.1186/s12302-015-0048-3.

Levin, Sam. 2018. "The Man Who Beat Monsanto: 'They Have to Pay for Not Being Honest.'" *The Guardian*, September 26, 2018. https://www.theguardian.com/business/2018/sep/25/monsanto-dewayne-johnson-cancer-verdict.

Liboiron, Max. 2021. *Pollution Is Colonialism*. Durham, NC: Duke University Press.

Liboiron, Max, Manuel Tironi, and Nerea Calvillo. 2017. "Toxic Politics: Acting in a Permanently Polluted World." *Social Studies of Science* 48 (3): 340–41.

Lichtman, J. S., J. A. Ferreyra, K. M. Ng, S. A. Smits, J. L. Sonnenburg, and J. E. Elias. 2016. "Host-Microbiota Interactions in the Pathogenesis of Antibiotic-Associated Diseases." *Cell Reports* 14 (5): 1049–61.

Lo, Puck. 2013. "Monsanto Bullies Famers over Planting Harvested GMO Seeds." *CorpWatch* (blog), March 24, 2013. https://corpwatch.org/article/monsanto-bullies-small-farmers-over-planting-harvested-gmo-seeds.

155

Los Angeles Times. 2010. "A History of Saccharin." December 27, 2010. https://www
.latimes.com/archives/la-xpm-2010-dec-27-la-he-nutrition-lab-saccharin
-timelin20101227-story.html.

Lyons, Kristina. 2018. "Chemical Warfare in Colombia, Evidentiary Ecologies
and Senti-Actuando Practices of Justice." *Social Studies of Science* 48 (3):
414–37.

MacLeash, Kenneth, and Zoe H. Wool. 2018. "US Military Burn Pits and the Politics of Health." *Critical Care* (blog), *Medical Anthropology Quarterly*, August 1,
2018. https://medanthroquarterly.org/critical-care/2018/08/us-military-burn
-pits-and-the-politics-of-health/.

Magaña-Gómez, Javier A., and Ana M. Calderón de la Barca. 2009. "Risk Assessment of Genetically Modified Crops for Nutrition and Health." *Nutrition
Review* 67 (1): 1–16.

Malatesta, Manuela, M. Biggiogera, E. Manuali, M. B. Rocchi, B. Baldelli, and
G. Gazzanelli. 2003. "Fine Structural Analyses of Pancreatic Acinar Cell
Nuclei from Mice Fed on Genetically Modified Soybean." *European Journal
of Histochemistry* 47 (4): 385–88.

Malatesta, Manuela, Federica Boraldi, Giulia Annovi, Beatrice Baldelli, Serafina
Battistelli, Marco Biggiogera, and Daniela Quaglino. 2008. "A Long-Term
Study on Female Mice Fed on a Genetically Modified Soybean: Effects on
Liver Ageing." *Histochemistry and Cell Biology* 130 (5): 967–77.

Malatesta, Manuela, Chiara Caporaloni, Stefano Gavaudan, Marco B. L. Rocchi,
Sonja Serafini, Cinzia Tiberi, and Giancarlo Gazzanelli. 2002. "Ultrastruc-
tural Morphometrical and Immunocytochemical Analyses of Hepatocyte
Nuclei from Mice Fed on Genetically Modified Soybean." *Cell Structure and
Function* 27: 173–80.

Malkan, Stacy. 2018. "Monsanto's Fingerprints All Over Newsweek's Opinion
Piece, a Hit on Organic Food." *EcoWatch*, January 14, 2018. https://www
.ecowatch.com/monsanto-propaganda-newsweek-2528277875.

Marabito, Maria (originally written by Jaymi Heimbuch). 2021 (2019). "The
Incredible Science behind Starling Murmurations: Where and Why They
Form." Treehugger, November 4, 2021. https://www.treehugger.com/the
-incredible-science-behind-starling-murmurations-4863751.

Marcus, George. 1995. "Ethnography in/of the World System: The Emergence of
Multi-Sited Ethnography." *Annual Review of Anthropology* 24: 95–117.

Markman, Shai, Carsten T. Muller, David Pascoe, Alistair Dawson, and Kath-
erine L. Buchanan. 2011. "Pollutants Affect Development in Nestling
Starlings *Sturnus vulgaris.*" *Journal of Applied Ecology* 48 (2): 391–97.

Martin, A., N. Myers, and A. Viseu. 2015. "The Politics of Care in Technoscience." *Social Studies of Science* 45 (5): 625–41.

Martineau, Belinda. 2001. *First Fruit: The Creation of the Flavr Savr™ Tomato and the
Birth of Biotech Food.* New York: McGraw-Hill.

Masco, Joseph. 2006. *The Nuclear Borderlands: The Manhattan Project in Post–Cold
War New Mexico.* Princeton, NJ: Princeton University Press.

156

Matthews, Jonathan. 2012. "Smelling a Corporate Rat." *Spin Watch*, December 11, 2012. http://www.spinwatch.org/index.php/issues/science/item/164 -smelling-a-corporate-rat.

McHenry, Leemon B. 2018. "The Monsanto Papers: Poisoning the Scientific Well." *International Journal of Risk and Safety in Medicine* 29 (3–4): 193–205.

Mesnage, Robin, Sarah Z. Agapito-Tenfen, Vinicius Vilperte, George Renney, Malcolm Ward, Gilles-Eric Séralini, Rubens O. Nodari, and Michael N. Antoniou. 2016. "An Integrated Multi-Omics Analysis of the NK603 Roundup-Tolerant GM Maize Reveals Metabolism Disturbances Caused by the Transformation Process." *Scientific Reports* 6: 37855.

Mesnage, Robin, and M. Antoniou. 2017. "Facts and Fallacies in the Debate on Glyphosate Toxicity." *Frontiers in Public Health*, November 24, 2017. https://www.frontiersin.org/articles/10.3389/fpubh.2017.00316/full.

Mesnage, Robin, Matthew Arno, Manuela Costanzo, Manuela Malatesta, Gilles-Eric Seralini, and Michael Antoniou. 2015. "Transcriptome Profile Analysis Reflects Rat Liver and Kidney Damage Following Chronic Ultra-Low Dose Roundup Exposure." *Environmental Health* 14 (70). https://ehjournal .biomedcentral.com/articles/10.1186/s12940-015-0056-1.

Mesnage, Robin, B. Bernay, and G. E. Séralini. 2013. "Ethoxylated Adjuvants of Glyphosate-Based Herbicides Are Active Principles of Human Cell Toxicity." *Toxicology* 313 (2–3): 122–28.

Mesnage, Robin, George Renney, Gilles-Eric Séralini, Malcolm Ward, and Michael Antoniou. 2017. "Multiomics Reveal Non-Alcoholic Fatty Liver Disease in Rats Following Chronic Exposure to an Ultra-Low Dose of Roundup Herbicide." *Scientific Reports* 7: 39328.

Miller, Daphne. 2013. *Farmacology: Total Health from the Ground Up*. New York: William Morrow.

Miller, Henry I. 2018. "The Campaign for Organic Foods Is a Deceitful, Expensive Scam." *Newsweek*, January 19, 2018. https://www.newsweek.com /campaign-organic-food-deceitful-expensive-scam-785493.

Mills, Paul J., Izabela Kania-Korwel, John Fagan, Linda K. McEvoy, Gail A. Laughlin, and Elizabeth Barrett-Connor. 2017. "Excretion of the Herbicide Glyphosate in Older Adults between 1993 and 2016." *JAMA* 318 (16): 1610–11.

Mol, Annemarie. 2004. *The Body Multiple: Ontologies in Medical Practice*. Durham, NC: Duke University Press.

Mol, Annemarie. 2008. "I Eat an Apple: On Theorizing Subjectivity." *Subjectivity* 22: 28–37.

Monsanto. 2017. "IARC's Report on Glyphosate." Monsanto (website), April 21, 2017. http://www.monsanto.com/iarc-roundup/pages/default.aspx.

Moschini, Gian Carlo. 2010. "Competition Issues in the Seed Industry and the Role of Intellectual Property." *Choices Magazine* 25 (2). https://dr.lib.iastate .edu/server/api/core/bitstreams/183b5a6d-47fc-4743-aff8-8f613af680fc /content.

157

Muller, Birgit. 2020. "Glyphosate—A Love Story: Ordinary Thoughtlessness and Response-Ability in Industrial Farming." *Journal of Agrarian Change* 21 (1): 160–79.

Muller, Martin. 2015. "Assemblages and Actor-Networks: Rethinking Socio-Material Power, Politics and Space." *Geography Compass* 9 (1): 27–41.

Murphy, Michelle. 2006. *Sick Building Syndrome and the Problem of Uncertainty*. Durham, NC: Duke University Press.

Murphy, Michelle. 2008. "Chemical Regimes of Living." *Environmental History* 13 (4): 695–703.

Murphy, Michelle. 2015. "Unsettling Care: Troubling Transnational Itineraries of Care in Feminist Health Practices." *Social Studies of Science* 45 (5): 717–37.

Murphy, Michelle. 2017a. *The Economization of Life*. Durham, NC: Duke University Press.

Murphy, Michelle. 2017b. "Alterlife and Decolonial Chemical Relations." *Cultural Anthropology* 32 (4): 494–503.

Murphy, Michelle. 2018a. "Against Population, towards Alterlife." In *Making Kin, Not Population*, edited by Adele Clarke and Donna Haraway, 101–24. Chicago: Prickly Paradigm Press.

Murphy, Michelle. 2018b. "Reimagining Chemicals, with and against Techno-science." In *Reactivating Elements*, edited by Dimitris Papadopoulos, María Puig de la Bellacasa, and Natasha Myers, 257–79. Durham, NC: Duke University Press.

Murphy, M. 2018c. "Gaslighting, Settler Colonialism, and Data Justice." Speech presented at Opaque Media: A Workshop, University of California Irvine, April 6, 2018. https://opaquemediaworkshop.wordpress.com/2018/02/05/announcement-conference-schedule/.

Myers, J. P., Michael N. Antoniou, Bruce Blumberg, Lynn Carroll, Theo Colborn, Lorne G. Everett, Michael Hansen, et al. 2016. "Concerns over Use of Glyphosate-Based Herbicides and Risks Associated with Exposures: A Consensus Statement." *Environmental Health* 15 (19). https://ehjournal.biomedcentral.com/articles/10.1186/s12940-016-0117-0.

Nading, Alex. 2015. "The Lively Ethics of GMOs: The Case of the Oxitec Mosquito." *Biosocieties* 10: 24–47.

Nading, Alex. 2016. "Evidentiary Symbiosis: On Paraethnography in Human-Microbe Relations." *Science as Culture* 25 (4): 560–81.

NASEM (National Academies of Science, Engineering and Medicine). 2016. *Genetically Engineered Crops: Experiences and Prospects*. Washington, DC: National Academies Press.

NASEM (National Academies of Sciences, Engineering, and Medicine). 2018. *Environmental Chemicals, the Human Microbiome, and Health Risk: A Research Strategy*. Washington, DC: National Academies Press. www.http://doi.org/10.17226/24960.

Nestle, Marion. 2018. *Unsavory Truth: How Food Companies Skew the Science of What We Eat*. New York: Basic Books.

NGIN (Norfolk Genetic Information Network). 2002. "JIC Man Key Player in Chapela Attacks—Nature on the Chapela Row." February 28, 2002, https://ngin.tripod.com/280202a.htm.

Nicolia, Alessandro, Alberto Manzo, Fabio Veronesi, and Daniele Rosellini. 2013. "An Overview of the Last 10 Years of Genetically Engineered Crop Safety Research." *Critical Reviews in Biotechnology* 34 (1): 1–12.

Niemann, Lars, Christian Sieke, Rudolf Pfeil, and Roland Solecki. 2015. "A Critical Review of Glyphosate Findings in Human Urine Samples and Comparison with the Exposure of Operators and Consumers." *Journal für Verbraucherschutz und Lebensmittelsicherheit* 10 (1): 3–12.

Oberlander, Herbert. 1993. "Howard A. Schneiderman: Planetary Patriot." *American Zoologist* 33 (3): 308–15.

OEHHA (California Office of Environmental Health Hazard Assessment). 2017a. "Glyphosate to Be Listed under Proposition 65 as Known to the State to Cause Cancer." OEHHA, March 28, 2017. https://oehha.ca.gov/proposition-65/crnr/glyphosate-be-listed-under-proposition-65-known-state-cause-cancer.

OEHHA (California Office of Environmental Health Hazard Assessment). 2017b. "Final Statement of Reasons, Title 27, California Code of Regulations, Section 2570(b) Specific Regulatory Levels Posing No Significant Risk, No Significant Risk Level: Glyphosate." https://oehha.ca.gov/media/downloads/crnr/glyphosatensrlfsor041018.pdf.

Ofrias, Lindsay. 2017. "Invisible Harms, Invisible Profits: A Theory of the Incentive to Contaminate." *Culture, Theory and Critique* 58 (4): 435–56.

Oreskes, Naomi, and Erik M. Conway. 2011. *Merchants of Doubt: How a Handful of Scientists Obscured the Truth on Issues from Tobacco Smoke to Global Warming.* London: Bloomsbury.

Paganelli, Alejandra, Victoria Gnazzo, Helena Acosta, Silvia L. López, and Andrés E. Carrasco. 2010. "Glyphosate-Based Herbicides Produce Teratogenic Effects on Vertebrates by Impairing Retinoic Acid Signaling." *Chemical Research in Toxicology* 23 (10) : 1586–95. doi: 10.1021/tx1001749. Epub 2010 Aug 9. PMID: 20695457.

Perro, Michelle, and Vincanne Adams. 2017. *What's Making Our Children Sick?* White River Junction, VT: Chelsea Green.

Pinch, Trevor J., and Wiebe E. Bijker. 1984. "The Social Construction of Facts and Artefacts, or How the Sociology of Science and the Sociology of Technology Might Benefit Each Other." *Social Studies of Science* 14 (3): 399–441.

Plumer, Brad. 2015. "Poll: Scientists Overwhelmingly Think GMOs Are Safe to Eat. The Public Doesn't." *Vox,* January 29, 2015. https://www.vox.com/2015/1/29/7947695/gmos-safety-poll.

Portier, C. J., Bruce K. Armstrong, Bruce C. Baguley, Zaver Bauer, Igor Belyaev, Robert Belle, Fiorella Belpoggi, et al. 2016. "Differences in the Carcinogenic Evaluation of Glyphosate between the International Agency for Research on Cancer (IARC) and the European Food Safety Authority (EFSA)." *Journal of Epidemiology in Community Health* 70 (8): 741–45.

159

Povinelli, Elizabeth A. 2016. *Geontologies: A Requiem for Late Liberalism*. Durham, NC: Duke University Press.

Povinelli, Elizabeth A. 2020. "The Virus: Figure and Infrastructure." *e-flux Architecture*, November. https://www.e-flux.com/architecture/sick-architecture /352870/the-virus-figure-and-infrastructure/.

Proctor, Robert N., and Londa Schiebinger. 2008. *Agnotology: The Making and Unmaking of Ignorance*. Stanford, CA: Stanford University Press.

Pryor, Lisa. 2018. "How to Counter the Circus of Pseudoscience." *New York Times*, January 5, 2018. https://www.nytimes.com/2018/01/05/opinion/doctors -naturopaths-health-science.html.

Puig de la Bellacasa, María. 2011. "Matters of Care in Technoscience: Assembling Neglected Things." *Social Studies of Science* 41 (1): 85–106.

Puig de la Bellacasa, María. 2017. *Matters of Care: Speculative Ethics in More Than Human Worlds*. Minneapolis: University of Minnesota Press.

Randerson, James. 2008. "Arpad Pustzai: Biological Divide." *Guardian*, January 15, 2008. https://www.theguardian.com/education/2008/jan/15 /academicexperts.highereducationprofile.

Rangel, Gabriel. 2015. "From Corgis to Corn: A Brief Look at the Long History of GMO Technology." *Science in the News* (blog), August 9, 2015. http://sitn .hms.harvard.edu/flash/2015/from-corgis-to-corn-a-brief-look-at-the-long -history-of-gmo-technology/.

Reese, Ashante M. 2019. *Black Food Geographies: Race, Self-Reliance, and Food Access in Washington, D.C.* Chapel Hill: University of North Carolina Press.

Relyea, Rick A. 2005. "The Lethal Impacts of Roundup and Predatory Stress on Six Species of North American Tadpoles." *Archives of Environmental Contamination and Toxicology* 48 (3): 351–57.

Reuters. 2009. "Corrected: Timeline: History of Monsanto Co." November 10, 2009. https://www.reuters.com/article/us-food-monsanto/corrected -timeline-history-of-monsanto-co-idUSTRE5AA05Q20091111.

Rhodes, Jonathan M. 1999. "Genetically Modified Foods and the Pusztai Affair." *BMJ* 318 (7193): 1284.

Richard, Sophie, Safa Moslemi, Herbert Sipahutar, Nora Benachour, and Gilles-Eric Séralini. 2005. "Differential Effects of Glyphosate and Roundup on Human Placental Cells and Aromatase." *Environmental Health Perspectives* 113 (6): 716–20.

Ricroch, Agnes E. 2013. "Assessment of GE Food Safety Using '-omics' Techniques and Long-Term Animal Feeding Studies" *Newe Biotechnology* 30 (4): 349–54.

Roberts, Jody A. 2014. "Unruly Technologies and Fractured Oversight: Toward a Model for Chemical Control for the Twenty-First Century." In Boudia and Jas, *Powerless Science? Science and Politics in a Toxic World*, vol. 2, chap. 12, loc. 5450.

Robin, Marie-Monique, dir. 2008. *The World According to Monsanto* (*Le monde selon Monsanto*). Documentary film, 1 hour, 49 minutes, 1 second. Image et Compagnie; Arte France; Office national du film du Canada; Productions

Thalie; Westdeutscher Rundfunk. https://www.youtube.com/watch?v
=hooBWyZHQ5Y.

Robin, Marie-Monique. 2010. *The World According to Monsanto: Pollution, Corruption, and the Control of the World's Food Supply*. New York: New Press.

Robinson, Claire, Michael Antoniou, and John Fagan. 2015. GMO *Myths and Truths: A Citizen's Guide to the Evidence on the Safety and Efficacy of Genetically Modified Crops and Foods*. 3rd ed. London: Earth Open Source.

Robinson, Claire, and Jonathan Latham. 2013. "The Goodman Affair: Monsanto Targets the Heart of Science." *Independent Science News*, May 20, 2013. https://www.independentsciencenews.org/science-media/the-goodman
-affair-monsanto-targets-the-heart-of-science.

Rowell, Andrew. 2003. *Don't Worry, It's Safe to Eat: The True Story of GM Food, BSE, and Foot and Mouth*. Oxford, UK: Earthscan Press.

Royal Society. 2016. "What GMO Crops Are Currently Being Grown and Where?" Accessed August 10, 2022. https://royalsociety.org/topics-policy/projects
/gm-plants/what-gm-crops-are-currently-being-grown-and-where/.

Sanchez Barbra, Mayra G. 2020. "'Keeping Them Down': Neurotoxic Pesticides, Race, and Disabling Biopolitics." *Catalyst: Feminism, Theory, Technoscience* 6 (1): 1–3.

Saxton, Dvera I. 2015. "Strawberry Fields as Extreme Environments: The Ecobiopolitics of Farmworker Health." *Medical Anthropology* 34 (2): 166–83.

Schell, J., and M. Van Montagu. 1977. "The Ti-Plasmid of *Agrobacterium tumefaciens*, a Natural Vector for the Introduction of NIF Genes in Plants?" In *Genetic Engineering for Nitrogen Fixation*, edited by A. Hollaender, R. H. Burris, P. R. Day, R. W. Hardy, D. R. Helsinkin, M. R. Lamborg, L. Owens, and R. C. Valentine, 159–79. New York: Plenum.

Schneider, Keith. 1990. "Betting the Farm on Biotech." *New York Times Magazine*, June 10, 1990. https://www.nytimes.com/1990/06/10/magazine/betting-the
-farm-on-biotech.html.

Schneiderman, Howard A. 1994. *Howard A. Schneiderman, a Brief Autobiography, 1927–1990*. Edited by Philip Needleman and Karen Keeler Rogers. St. Louis, MO: Monsanto.

Schneiderman, Howard A., and Will Carpenter. 1990. "Planetary Patriotism: Sustainable Agriculture for the Future." *Environmental Science and Technology* 24 (4): 466–73.

Schuette, J. 1998. *Environmental Fate of Glyphosate*. White Paper. Sacramento, CA: Environmental Monitoring and Pest Management, Department of Pesticide Regulation, Sacramento. https://silo.tips/download/environmental
-fate-of-glyphosate-jeff-schuette.

Schurman, Rachel, and William A. Munro. 2010. *Fighting for the Future of Food: Activists versus Agribusiness in the Struggle over Biotechnology*. Minneapolis: University of Minnesota Press.

Seligman, Hilary, Terry C. Davis, Dean Schillinger, and Michael S. Wolf. 2010. "Food Insecurity Is Associated with Hypoglycemia and Poor Diabetes

Self-Management in a Low-Income Sample with Diabetes." *Journal of Health Care for the Poor and Underserved* 21 (4): 1227–33.

Séralini, Gilles-Eric, Dominique Cellier, and Joël Spiroux de Vendomois. 2007. "New Analysis of a Rat Feeding Study with a Genetically Modified Maize Reveals Signs of the Hepatorenal Toxicity." *Archives of Environmental Contamination and Toxicology* 52 (4): 596–602.

Séralini, Gilles-Eric, Emilie Clair, Robin Mesnage, Steeve Gress, Nicolas Defarge, Manuela Malatesta, Didier Hennequin, et al. 2012. "Retracted: Long-Term Toxicity of a Roundup Herbicide and a Roundup-Tolerant Genetically Modified Maize." *Food and Chemical Toxicology* 50 (11): 4221–31.

Séralini, Gilles-Eric, Emilie Clair, Robin Mesnage, Steeve Gress, Nicolas Defarge, Manuela Malatesta, Didier Hennequin, and Joel Spiroux de Vendomois. 2014a. "Republished Study: Long-Term Toxicity of a Roundup Herbicide and a Roundup-Tolerant Genetically Modified Maize." *Environmental Sciences Europe* 26 (14). https://enveurope.springeropen.com/articles/10.1186/s12302-014-0014-5.

Séralini, Gilles-Eric, R. Mesnage, N. Defarge, and J. Spiroux de Vendomois. 2014b. "Conclusiveness of Toxicity Data and Double Standards." *Food and Chemical Toxicology* 69: 357–59.

Shadaan, Reena, and Michelle Murphy. 2020. "Endocrine-Disrupting Chemicals (EDCs) as Industrial and Settler Colonial Structures: Towards a Decolonial Feminist Approach." *Catalyst: Feminism, Theory, Technoscience* 6 (1): 1–36.

Shapiro, Nicholas. 2015. "Attuning to the Chemosphere: Domestic Formaldehyde, Bodily Reasoning, and the Chemical Sublime." *Cultural Anthropology* 30 (3): 368–93.

Shapiro, Nicholas, and Eben Kirksey. 2017. "Chemo-Ethnography: An Introduction." *Cultural Anthropology* 32 (4): 481–93.

Shehata, Awad, Wieland Schrödl, Alaa A. Aldin, Hafez M. Hafez, and Monika Kruger. 2013. "The Effect of Glyphosate on Potential Pathogens and Beneficial Members of Poultry Microbiota In Vitro." *Current Microbiology* 66 (4): 350–58.

Sherman, Janette D. 1996. "Chlorpyrifos (Dursban)-Associated Birth Defects: Report of Four Cases." *Archives of Environmental Health* 51 (1): 5–8.

Shiva, Vandana. 2000. *Stolen Harvest: The Hijacking of the Global Food Supply.* Cambridge, MA: South End Press.

Simanis, Erik. 2001. *The Monsanto Corporation: A Quest for Sustainability (A).* Washington, DC: World Resources Institute. http://pdf.wri.org/bell/case_15-69734-755-_full_version_a_english.pdf.

Slaughter, Sheila, and Larry L. Leslie. 1997. *Academic Capitalism: Politics, Policies, and the Entrepreneurial University.* Baltimore, MD: Johns Hopkins University Press.

Smith, Michael B. 2001. "'Silence, Miss Carson!' Science, Gender, and the Reception of *Silent Spring*." *Feminist Studies* 27 (3): 733–52.

Snell, Chelsea, Aude Bernheim, Jean-Baptiste Bergé, Marcel Kuntz, Gérard Pascal, Alain Paris, and Agnès E. Ricroch. 2012. "Assessment of the Health Impact of GM Plant Diets in Long-Term and Multigenerational Animal

Feeding Trials: A Literature Review." *Food and Chemical Toxicology* 50 (3–4): 11344–8.

Sonnenberg, J. L., and F. Backhed. 2016. "Diet-Microbiota Interactions as Moderators of Human Metabolism." *Nature* 535 (7610): 56–64.

Spanogiannopoulos, Peter, Elizabeth N. Bess, Rachel N. Carmody, and Peter J. Turnbaugh. 2016. "The Microbial Pharmacists within Us: A Metagenomic View of Xenobiotic Metabolism." *Nature Reviews/Microbiology* 14: 273–87.

Star, Susan Leigh, and James R. Greisemer. 1989. "Institutional Ecology, 'Translations' and Boundary Objects: Amateurs and Professionals in Berkeley's Museum of Vertebrate Zoology, 1907–39." *Social Studies of Science* 19 (3): 387–420.

Starr, Paul. 1984. *The Social Transformation of American Medicine*. New York: Basic Books.

Stengers, Isabelle. 2003. "The Doctor and the Charlatan." *Cultural Studies Review* 9 (2): 11–36.

Stengers, Isabelle. 2018. *Another Science Is Possible: A Plea for Slow Science*. Translated by Stephen Muecke. Boston: Polity.

Stengers, Isabelle, and Peter Weibel. 2011. *Cosmopolitics II*. Translated by Robert Bononno. Minneapolis: University of Minnesota Press.

Steptoe. 2020. "Court Overturns Prop. 65 Warning for Glyphosate as Unconstitutional Compelled Speech: The Beginning of a Broad New Defense in Prop. 65 Cases?" Steptoe Client Alerts, June 25, 2020. https://www
.steptoe.com/en/news-publications/court-overturns-prop-65-warning-for
-glyphosate-as-unconstitutional-compelled-speech-the-beginning-of-a
-broad-new-defense-in-prop-65-cases.html.

Stiles, Michael. 2017. "New York Times Article Resembles Press Release." Medium, August 30, 2017. https://medium.com/@michaelpacificosur/new
-york-times-article-resembles-press-release-302b992e6bcb.

Stone, Glenn Davis. 2002. "Both Sides Now: Fallacies in the Genetic Modification Wars, Implications for Developing Countries, and Anthropological Perspectives." *Current Anthropology* 43 (4): 611–30.

Stone, Glenn Davis. 2010. "The Anthropology of Genetically Modified Crops." *Annual Review of Anthropology* 39: 381–400.

Stone, Glenn Davis. 2015. "Biotechnology, Schizmogenesis and the Demise of Uncertainty." *Journal of Law and Policy* 47: 29.

Stone, Glenn Davis. 2017. "Dreading CRISPR: GMOs, Honest Brokers, and Mertonian Transgressions." *Geographical Review* 107 (4): 1–8.

Stone, Glenn Davis. 2018. "The Dubious Virtue of Apostasy: How a Former Activist Saw the GMO Light and Is Being Amply Rewarded for It." Review of *Seeds of Science: Why We Got It So Wrong on GMOs*, by Mark Lynas. *Common Reader*, December 13, 2018. https://commonreader.wustl.edu/c/the-dubious
-virtue-of-apostasy/.

Strodder, Chris. 2017. *The Disneyland Encyclopedia*. Santa Monica, CA: Santa Monica Press.

Sullivan, Emily. 2018. "Groundskeeper Accepts Reduced $78 Million Award in Monsanto Cancer Suit." *National Public Radio*, November 1, 2018. https://

163

www.npr.org/2018/11/01/662812333/groundskeeper-accepts-reduced-78
-million-in-monsanto-cancer-suit.

Sutton, Patrice, David Wallinga, Joane Perron, Michelle Gottlieb, Lucia Sayre, and Tracey Woodruff. 2011. "Reproductive Health and the Industrialized Food System: A Point of Intervention for Health Policy." *Health Affairs* 30 (5): 888–97.

Székács, András, and Béla Darvas. 2013. "Comparative Aspects of Cry Toxin Usage in Insect Control." In *Advanced Technologies for Managing Insect Pests*, edited by Isaac Ishaaya, Subba Reddy Palli, and Rami Horowitz, 195–230. New York City: Springer.

Taylor, Sunaura. 2017. *Beasts of Burden: Animal and Disability Liberation*. New York: New Press.

Thongprakaisang, Siriporn, Apinya Thiantanawat, Nuchanart Rangkadilok, Tawit Suriyo, and Jutamaad Satayavivad. 2013. "Glyphosate Induces Human Breast Cancer Cell Growth via Estrogen Receptors." *Food and Chemical Toxicology* 59: 129–36.

Tickell, Josh. 2017. *Kiss the Ground: How the Food You Eat Can Reverse Climate Change, Heal Your Body and Ultimately Save Our World*. Miami, FL: Atria/Enliven Books.

Ticktin, Miriam. 2017. "A World without Innocence." *American Ethnologist* 44 (4): 577–90.

Tironi, Manuel. 2018. "Hypo-Interventions: Intimate Activism in Toxic Environments." *Social Studies of Science* 48 (3): 438–55.

Tokar, Brian. 1998. "Monsanto: A Checkered History." *Ecologist*, September 1, 1998. https://theecologist.org/1998/sep/01/monsanto-checkered-history.

Touray, Seringe S. T. 2019. "Cause of Mass Deaths of Starlings in the Hague Identified." *Holland Times*. February 1, 2019. https://www.hollandtimes.nl/articles/national/cause-of-mass-deaths-of-starlings-in-the-hague-identified/.

Tsing, Anna Lowenhaupt. 2015. *The Mushroom at the End of the World: On the Possibility of Life in Capitalist Ruins*. Princeton, NJ: Princeton University Press.

Tsing, Anna Lowenhaupt, Heather Anne Swanson, Elain Gan, and Nils Bubandt, eds. 2017. *Arts of Living on a Dying Planet*. Minneapolis: University of Minnesota Press.

US Right to Know. n.d.a "Have You Had Prolonged Exposure to Monsanto's Roundup? Act Now: Hundreds of Lawsuits Have Been Filed." Accessed March 12, 2021. https://roundup-cancer-claim.com/?utm_source=google&utm_term=%2Bnon%20%2Bhodgkin%27s%20%2Blymphoma%20%2Bmonsanto~c~g&gclid=CjoKCQiA3smABhCjARIsAKtrg6KPnB43SXtDVpba4TsU_wriADxfowVAUr4eYYns7g8oVWpRKFLpIsQaAn-IEALw_wcB.

US Right to Know. n.d.b "Roundup (Glyphosate) Cancer Cases: Key Documents and Analysis." Accessed July 21, 2021. https://usrtk.org/monsanto-papers/.

Vaeck, M., A. Reynaerts, H. Höfte, Stefan Jansens, Marc De Beuckeleer, Caroline Dean, Marc Zabeau, et al. 1987. "Transgenic Plants Protected from Insect Attack." *Nature* 328: 33–37.

Van Eenennaam, A. L., J. Li, R. M. Thallman, R. L. Quaas, M. E. Dikeman, C. A. Gill, D. E. Franke, and M. G. Thomas. 2007. "Validation of Commercial

164

DNA Tests for Quantitative Beef Quality Traits." *Journal of Animal Science* 85 (4): 891–900.

Van Eenennaam, A. L., and A. E. Young. 2014. "Prevalence and Impacts of Genetically Engineered Feedstuffs on Livestock Populations." *Journal of Animal Science* 92 (10): 4255–78.

Vecchio, L., B. Cisterna, M. Malatesta, T. E. Martin, and M. Biggiogera. 2004. "Ultrastructural Analysis of Testes from Mice Fed on Genetically Modified Soybean." *European Journal of Histochemistry* 48 (4): 448–54.

Walker, Brett L. 2011. *Toxic Archipelago: A History of Industrial Disease in Japan.* Seattle: University of Washington Press.

Walker, J. S. 2000. *Permissible Dose: A History of Radiation Protection in the Twentieth Century.* Berkeley: University of California Press.

Walsh, Maria C., Stefan G. Buzoianu, Gillian E. Gardiner, Mary C. Rea, Eva Gelencser, Anna Janosi, Michelle M. Epstein, et al. 2011. "Fate of Transgenic DNA from Orally Administered Bt MON810 Maize and Effects on Immune Response and Growth in Pigs." *PLOS One* 6 (11): e27177.

Weasel, Lisa H. 2009. *Food Fray: Inside the Controversy over Genetically Modified Food.* New York: AMACOM.

Weiss, Rick. 1999. "Seeds of Discord—Monsanto's Gene Police Raise Alarm on Farmers' Rights, Rural Tradition." *Washington Post*, February 3, 1999. https://www.washingtonpost.com/archive/politics/1999/02/03/seeds-of-discord/c0f613a0-02a1-476f-b54d-af25413844f5/.

Weston, Kath. 2017. *Animate Planet: Making Visceral Sense of Living in a High-Tech Ecologically Damaged World.* Durham, NC: Duke University Press

Wild, Christopher. 2018. "IARC Response to Criticisms of the Monographs and the Glyphosate Evaluation." International Agency for Research on Cancer, January 2017. https://iarc.who.int/wpcontent/uploads/2018/07/IARC_response_to_criticisms_of_the_Monographs_and_the_glyphosate_evaluation.pdf.

Williams, G. M., R. Kroes, and I. C. Munro. 2000. "Safety Evaluation and Risk Assessment of the Herbicide Roundup and Its Active Ingredient, Glyphosate, for Humans." *Regulatory Toxicology and Pharmacology* 31 (2): 117–65.

Woolgar, Steve, and Javier Lezaun. 2015. "The Wrong Bin Bag: A Turn to Ontology in Science and Technology Studies." *Social Studies of Science* 43 (3): 321–34.

Woolgar, Steve, and Daniel Neyland. 2014. *Mundane Governance: Ontology and Accountability.* Oxford: Oxford University Press.

Wurster, Charles F. 1968. "DDT and Robins." *Science* 159: 1413–14.

Wurster, Charles F. 2015. *DDT Wars: Rescuing Our National Bird, Preventing Cancer, and Creating the Environmental Defense Fund.* Oxford: Oxford University Press.

Zdziarski, Irena M., Judy A. Carman, and John W. Edwards. 2018. "Histopathological Investigation of the Stomach of Rats Fed a 60% Genetically Modified Corn Diet." *Food and Nutrition Sciences* 9: 763–96.

Zierler, David. 2011. *The Invention of Ecocide: Agent Orange, Vietnam, and the Scientists Who Changed the Way We Think about the Environment.* Athens: University of Georgia Press.

GE seed production market, 33
Glantz, Stanton, 127
glyphosate. *See specific topics*
glyphosate-resistant foods, 14
GMO generation, 3
GMO Myths and Truths (Robinson), 92
GNA (*Galanthus nivalis*), 88
Goldberg, Ray A., 134
Grant, Hugh, 134, 135
GRAS. *See* generally recognized as safe
Greisemer, James, 103
Guattari, Felix, 71

Hardeman, Edwin, 126, 127
harm. *See* chemical harm
health panics, child, 5
Hendlin, Yogi, 41, 49, 55
hepatorenal toxicity, 90
heteronormative demands, 5
Honeycutt, Zen, 119
Hoover Institute, 98
"How Did the US EPA and IARC Reach
    Diametrically Opposed Conclusions
    on the Genotoxicity of Glyphosate-
    Based Herbicides?," 116

IARC. *See* International Agency for
    Research on Cancer
IgE (anaphylactic responses), 60
imaginative chemistry, 28
immune system responses, 60
industrialism, agrochemical, 5, 8, 11, 13,
    37, 117
industry bias, 98
industry corruption, 56
industry-driven science, 100
industry funding, 99
interabsorptions, of glyphosate, 70
interestedness, 108
International Agency for Research on
    Cancer (IARC), 7, 16, 114–16, 118
intestinal inflammation, in pigs, 82
investments: agrocapitalist, 15; agro-
    chemical, 25
Irene (patient), 61–64
Italy, 84

Jaenisch, Rudolf, 139n3
Jain, Lochlann, 51
Japan, 12
Jas, Nathalie, 48, 74
Jasanoff, 55
J. D. Searle (pharmaceutical company),
    28
Johns Hopkins University, 18
Johnson, Dewayne, 124–26, 127
Johnson and Johnson Pharmaceuti-
    cals, 16
Jukes, Thomas, 21–22

Kimbrell, Andrew, 42
King's College London, 68, 92
Kirksey, Eben, 6, 12
knowledge: clustering of, 110; counter-
    knowledge production, 74
Kuhn, Thomas, 86

Lancet (journal), 89, 91
Landecker, Hannah, 12, 131
Lasso, 24
Latham, Jonathan, 47, 140n3
lawsuits, patent infringement, 41
leaky gut, 68
legal appeals, by Monsanto, 7
Leslie, Larry L., 99
Lezaun, Javier, 37
life sciences, 19
livestock, 82, 83
Livingston, Julie, 14
Li-Young Lee, 1

MADLs. *See* maximum allowable dose
    levels
Malthusian futurity, 25
mammalian cells, 119–20
Marcus, George, 10
marketing: messages, 25; of Roundup,
    34
Martin, Henri, 16
Martineau, Belinda, 94
maximum allowable dose levels
    (MADLs), 46
Melissa (patient), 59

171

172

173